ANATOMY COLORING BOOK

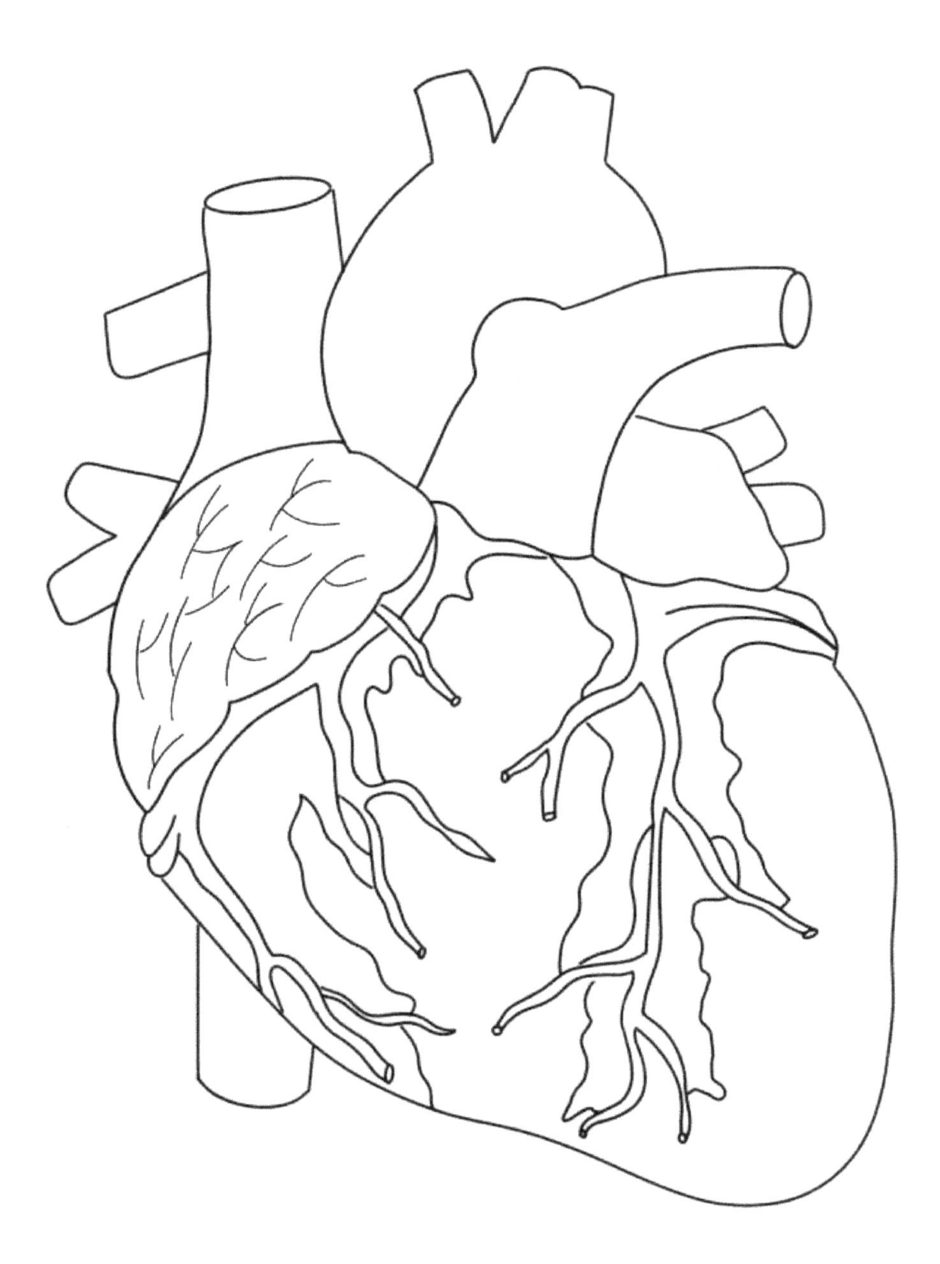

Copyright © 2020 by Bee Art Press
All rights reserved. This book or any part of it may not be reproduced or used in any way without the express written permission of the publisher, except for the use of brief quotations in a book review.

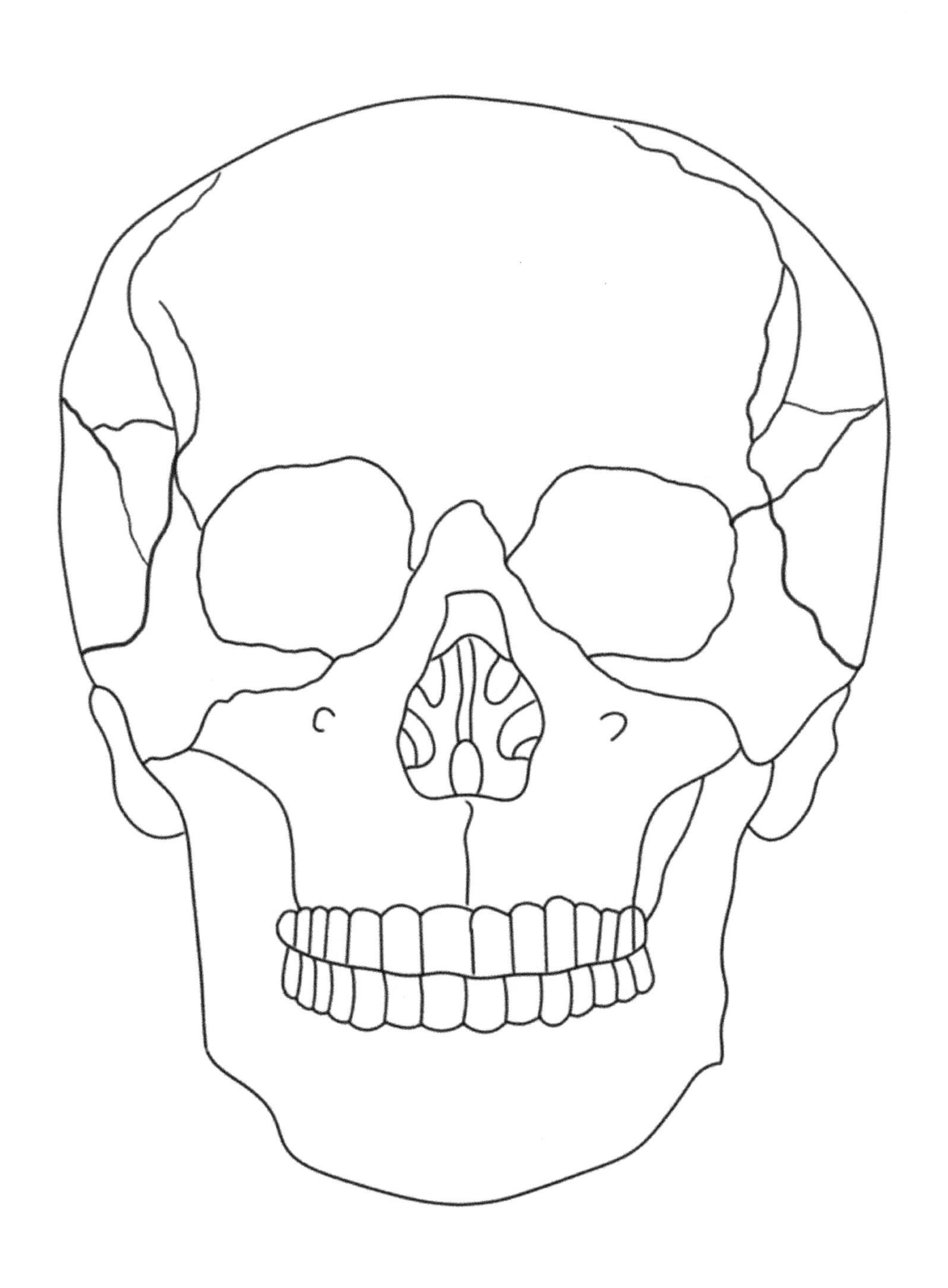

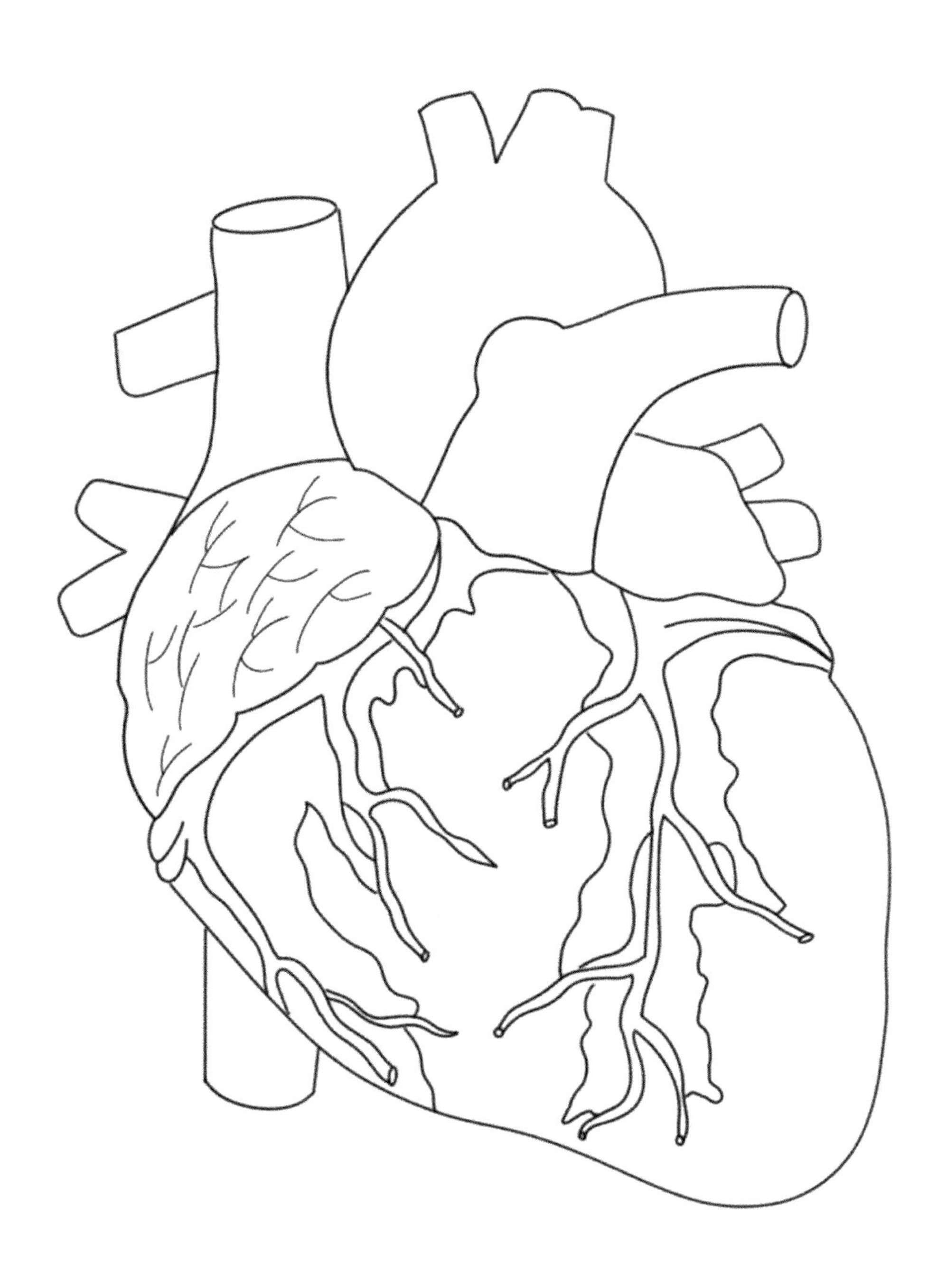

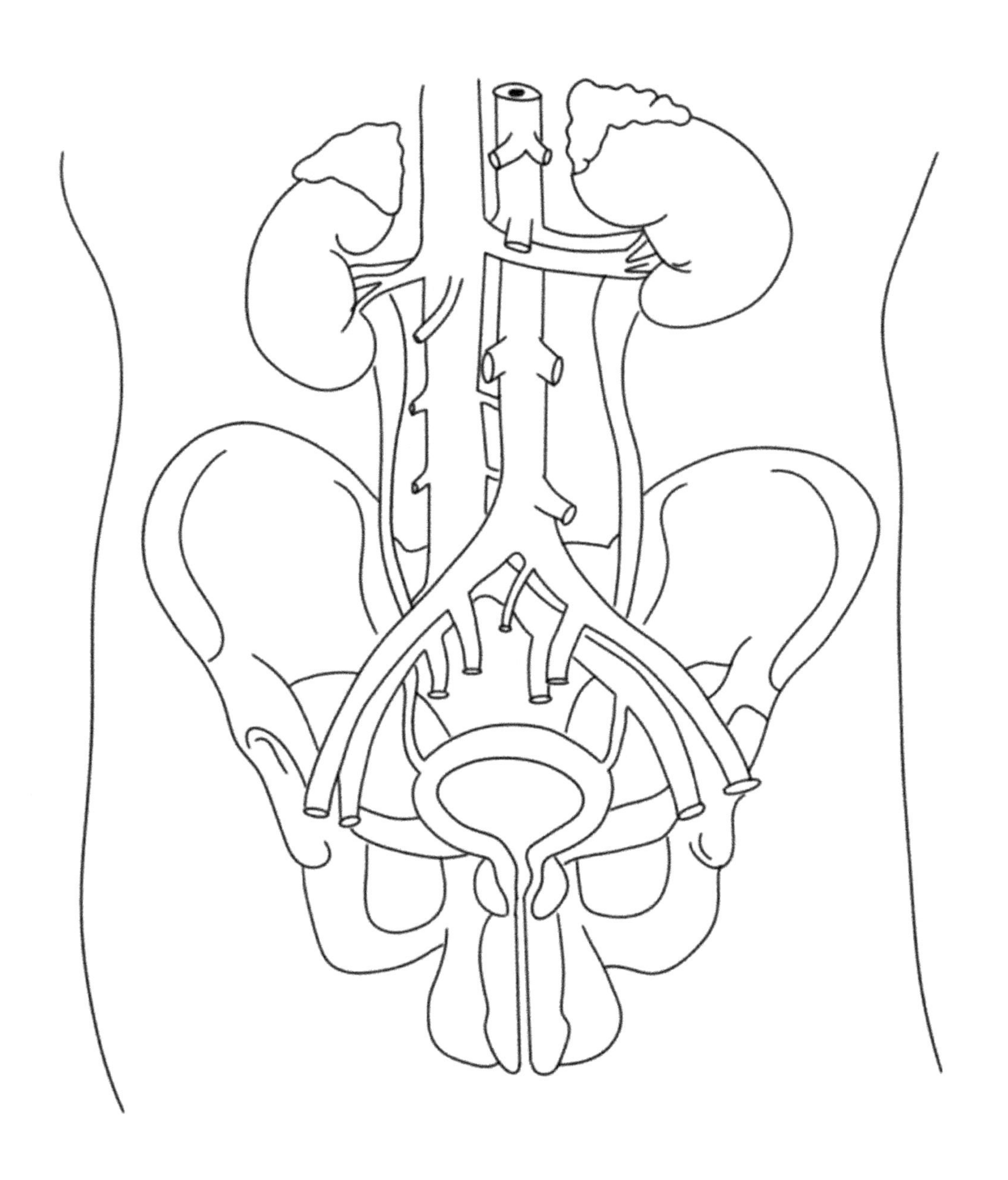

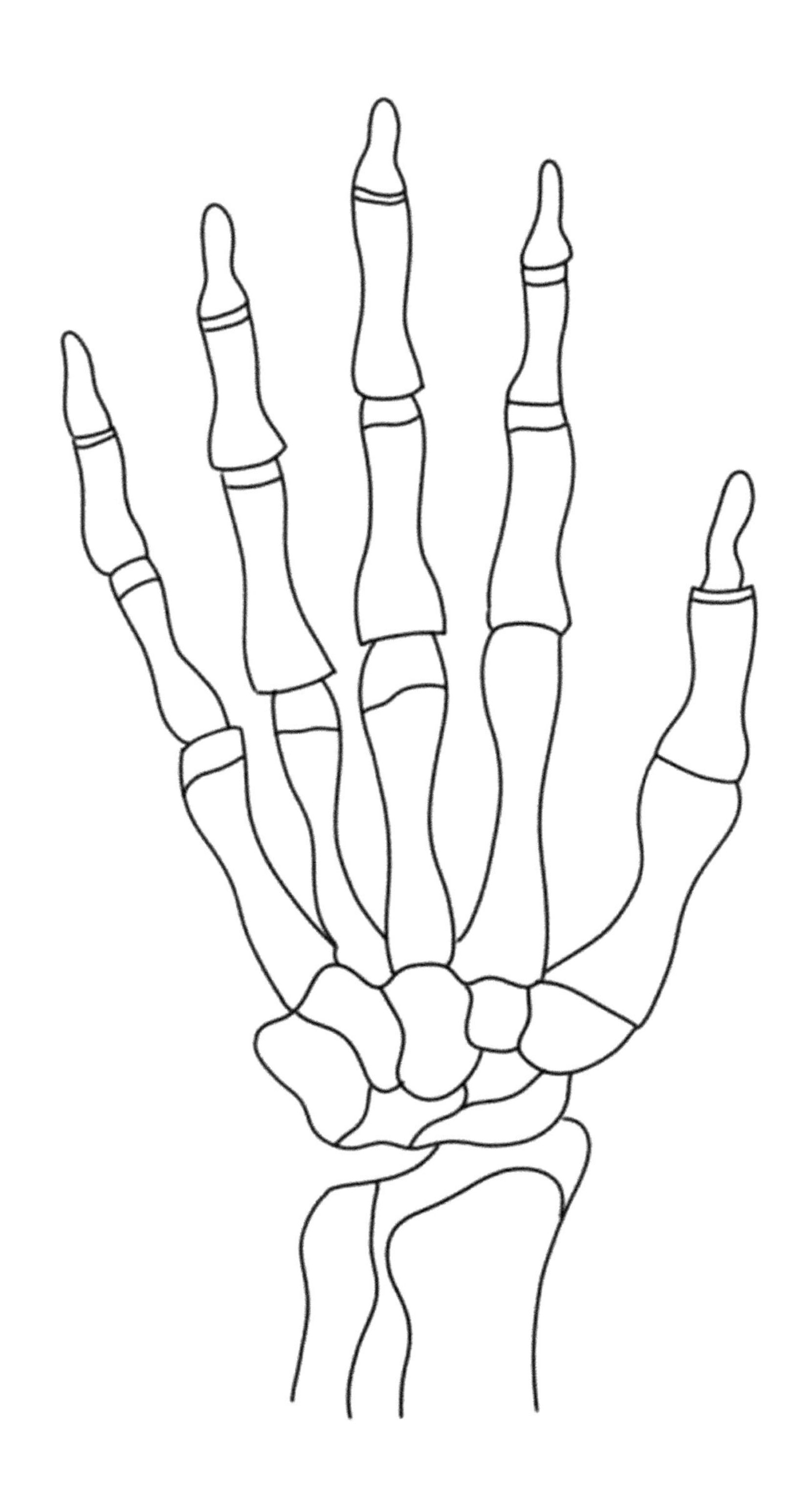

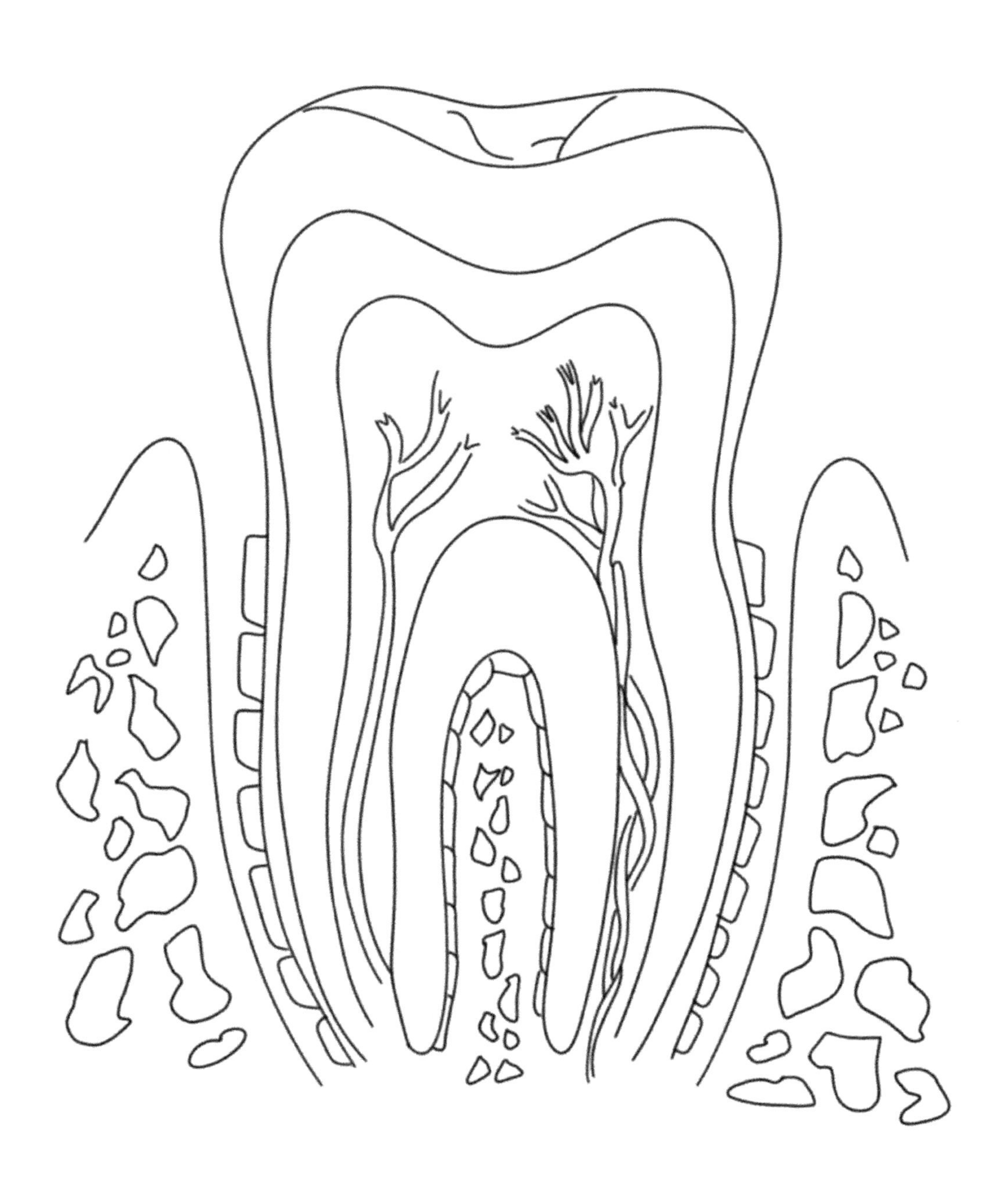

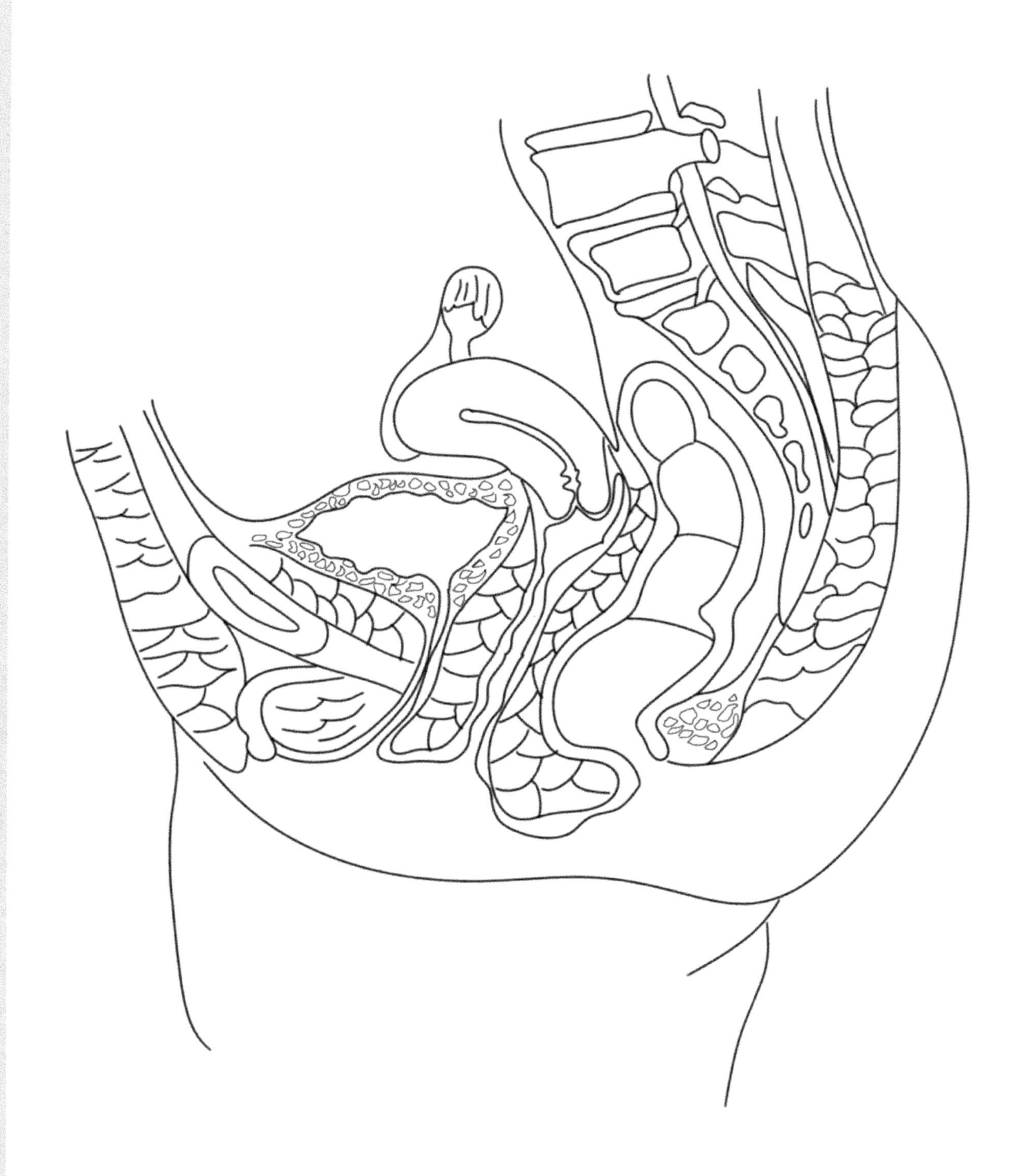

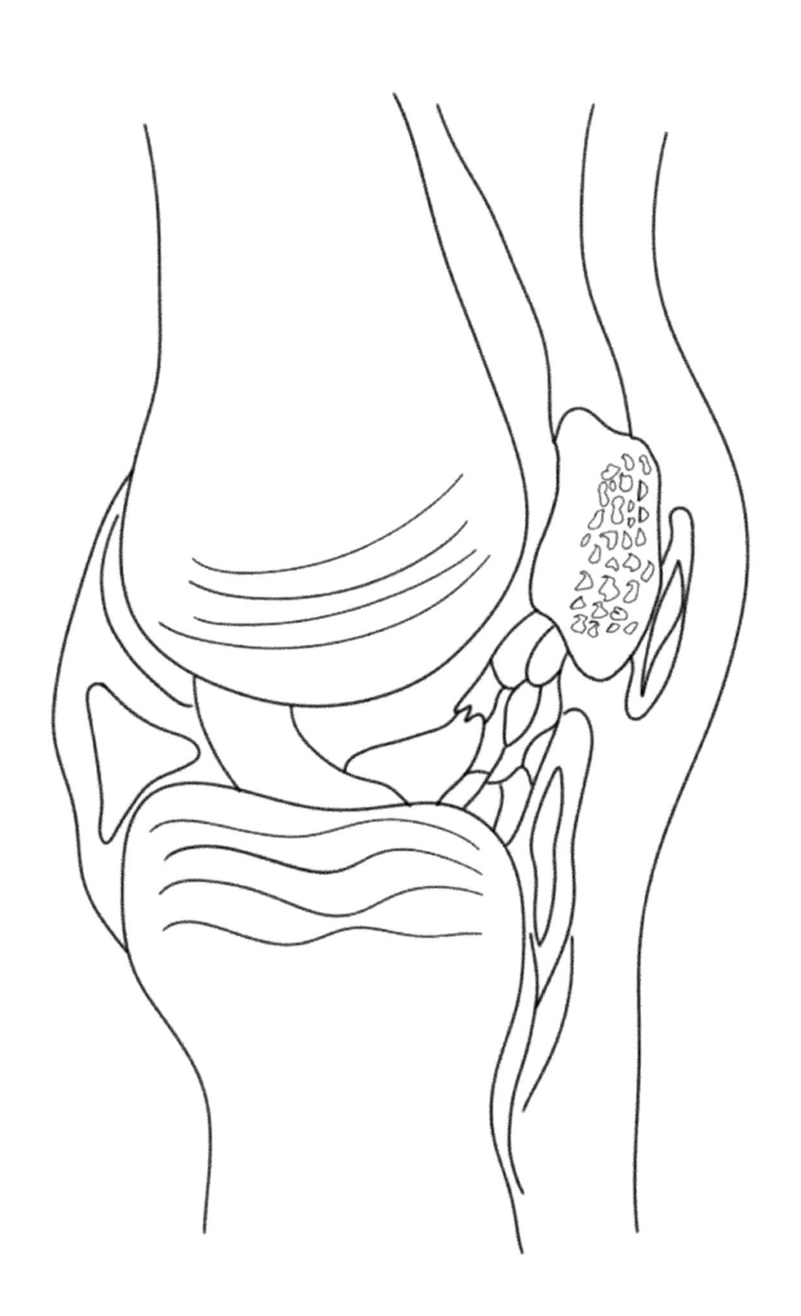

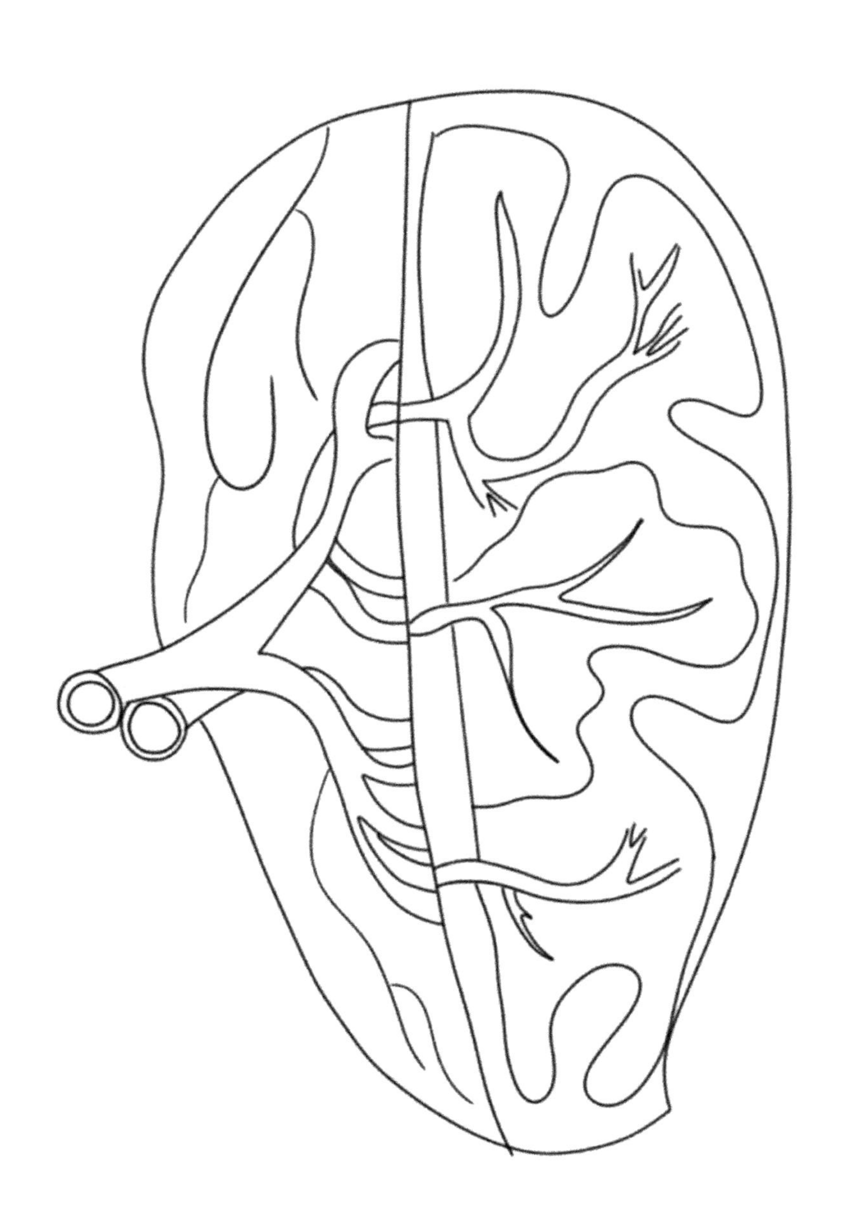

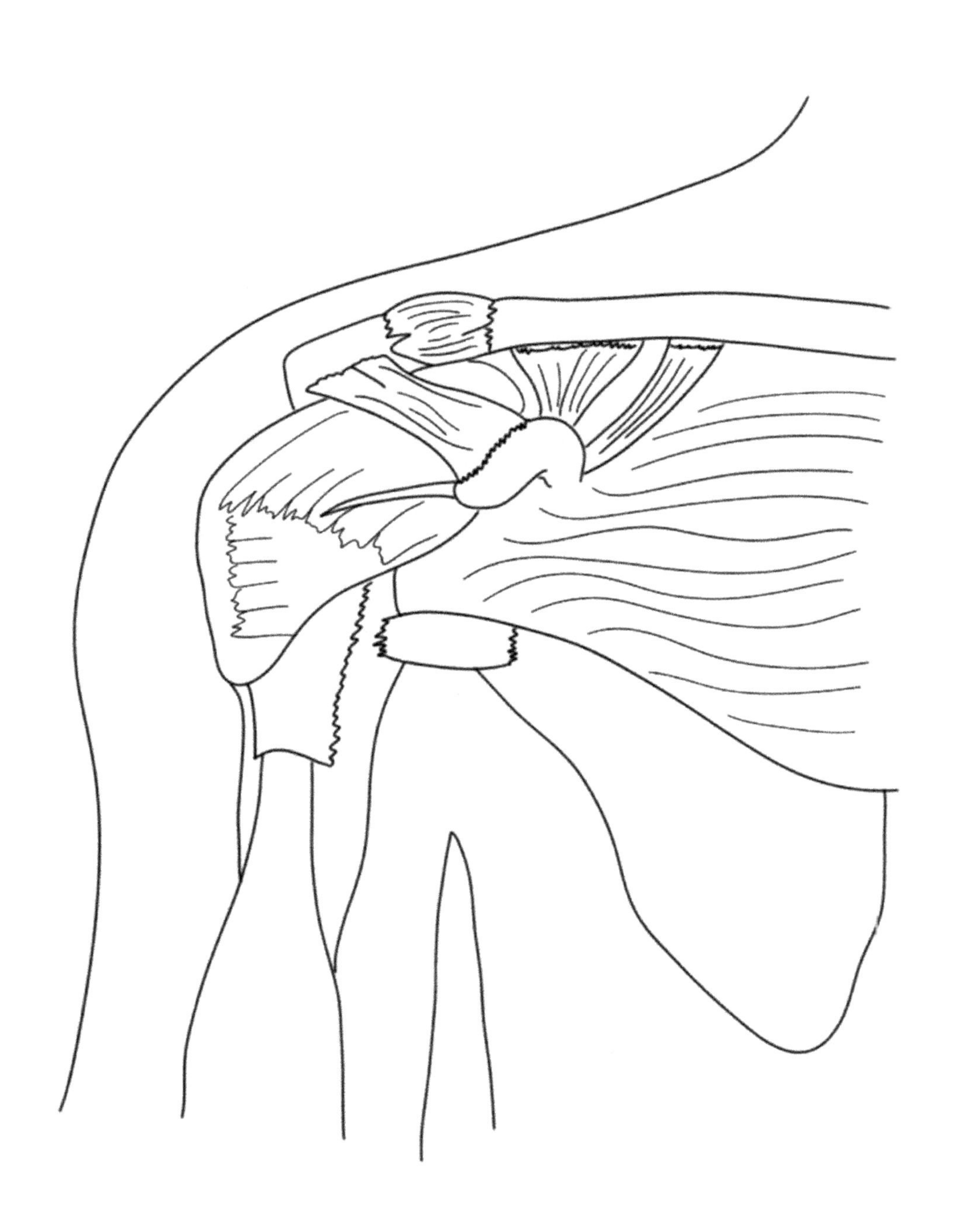

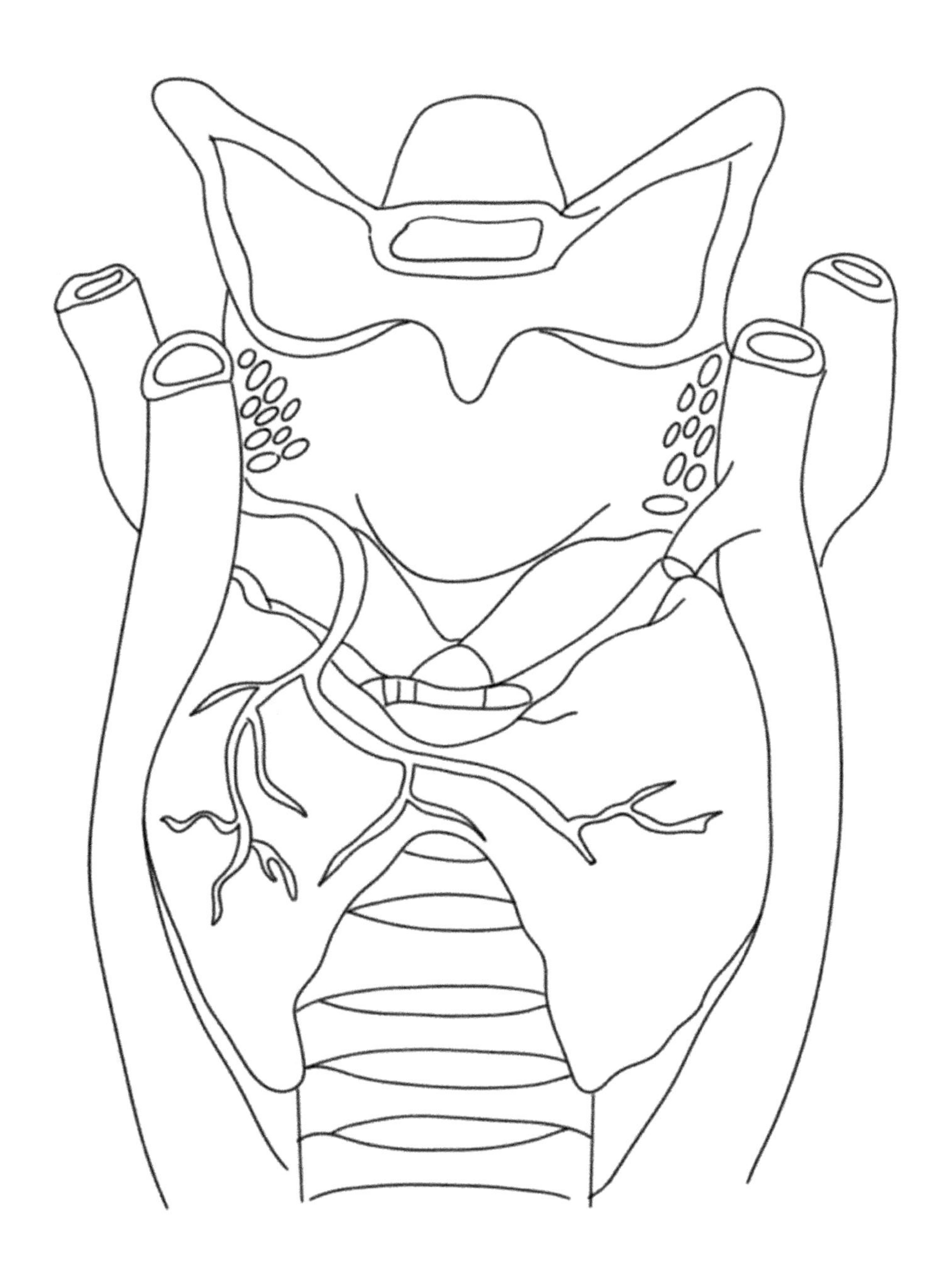

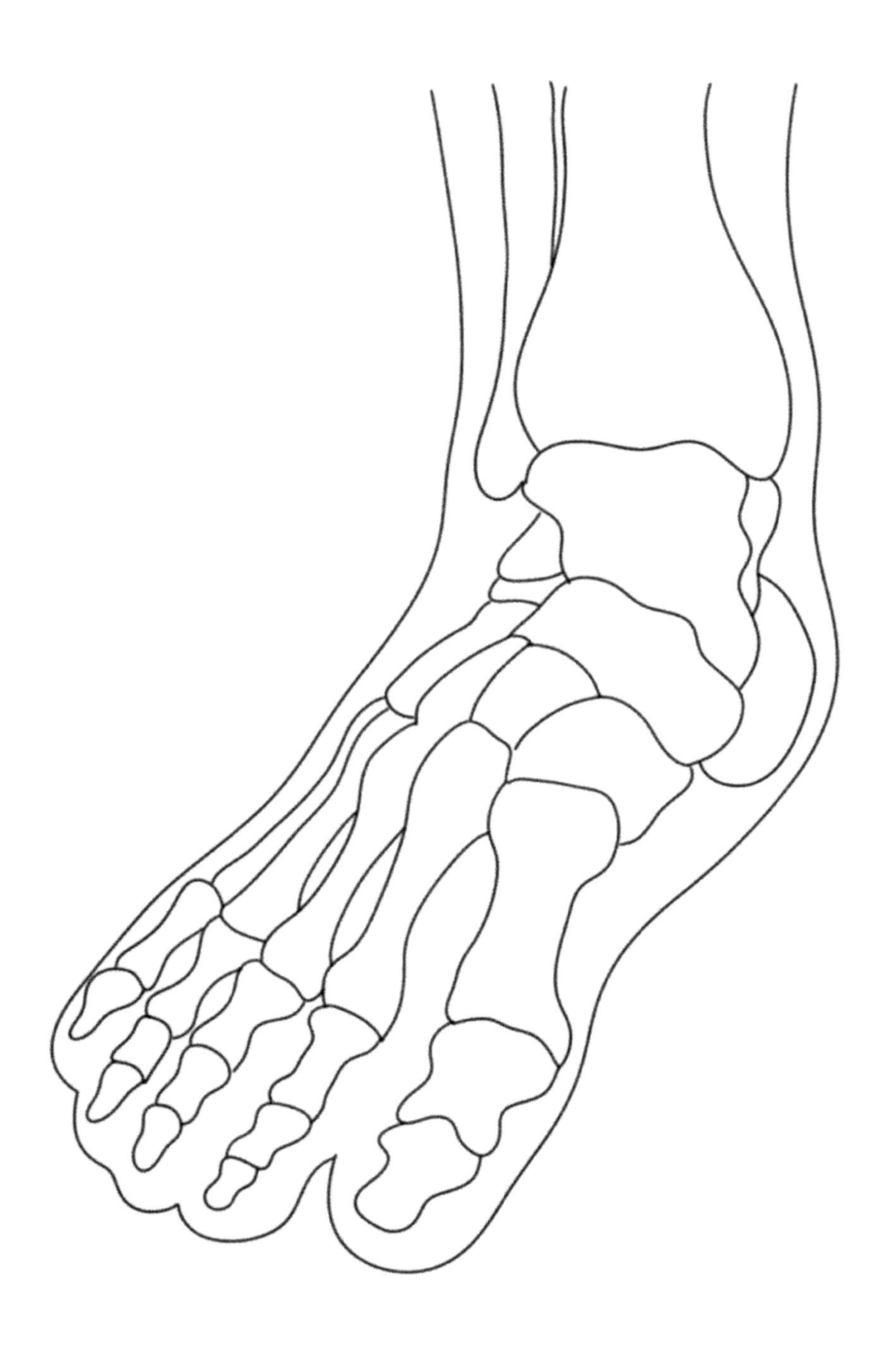

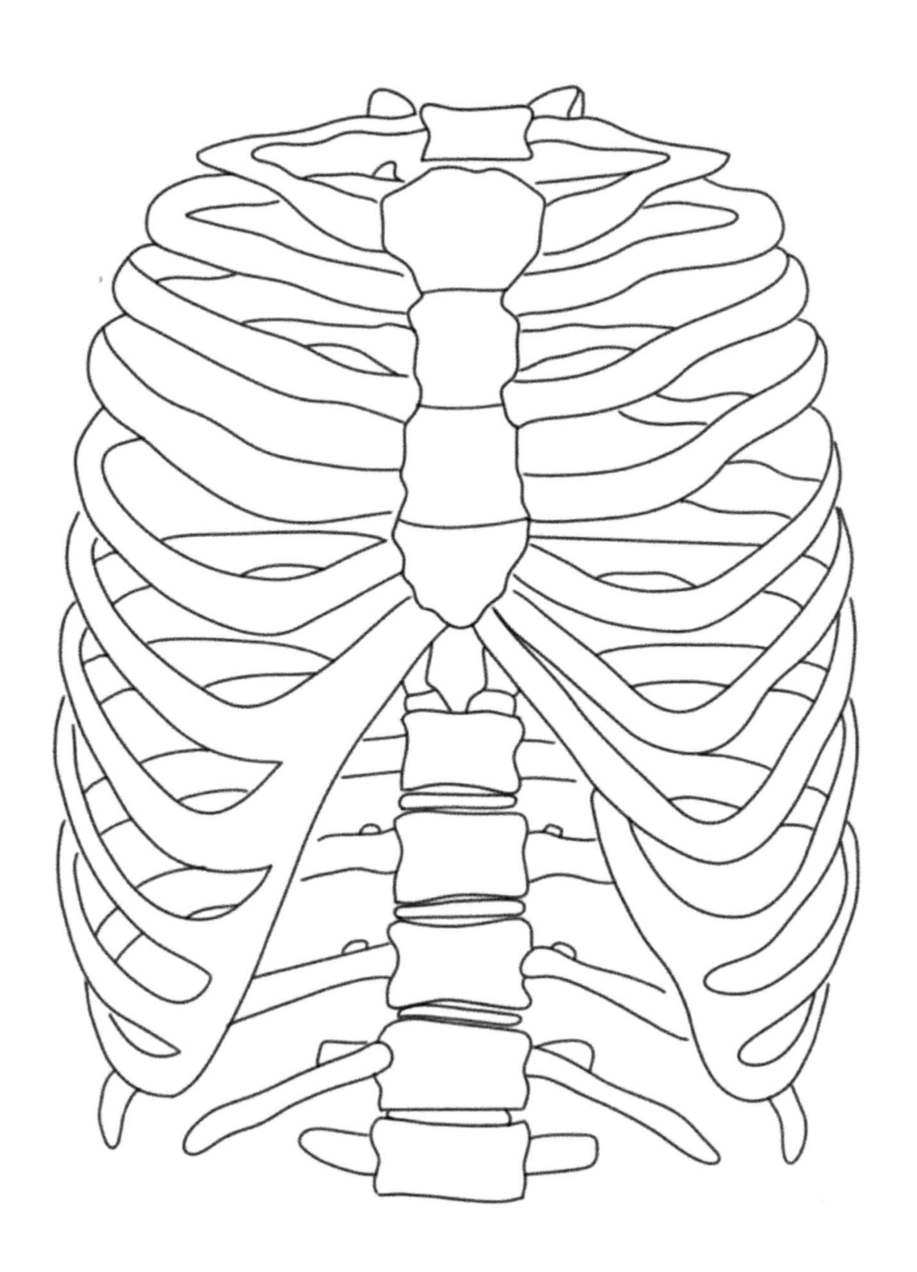

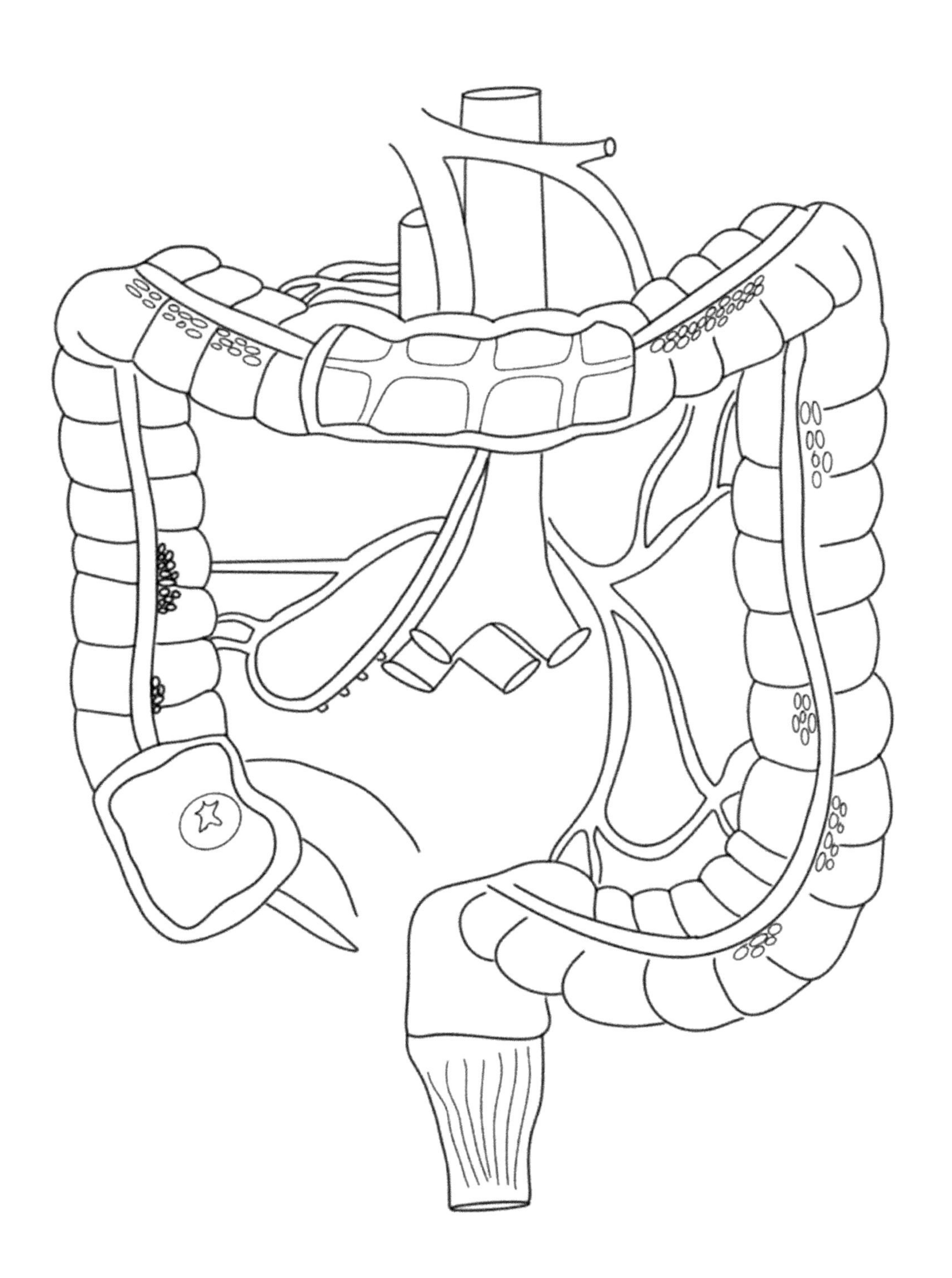

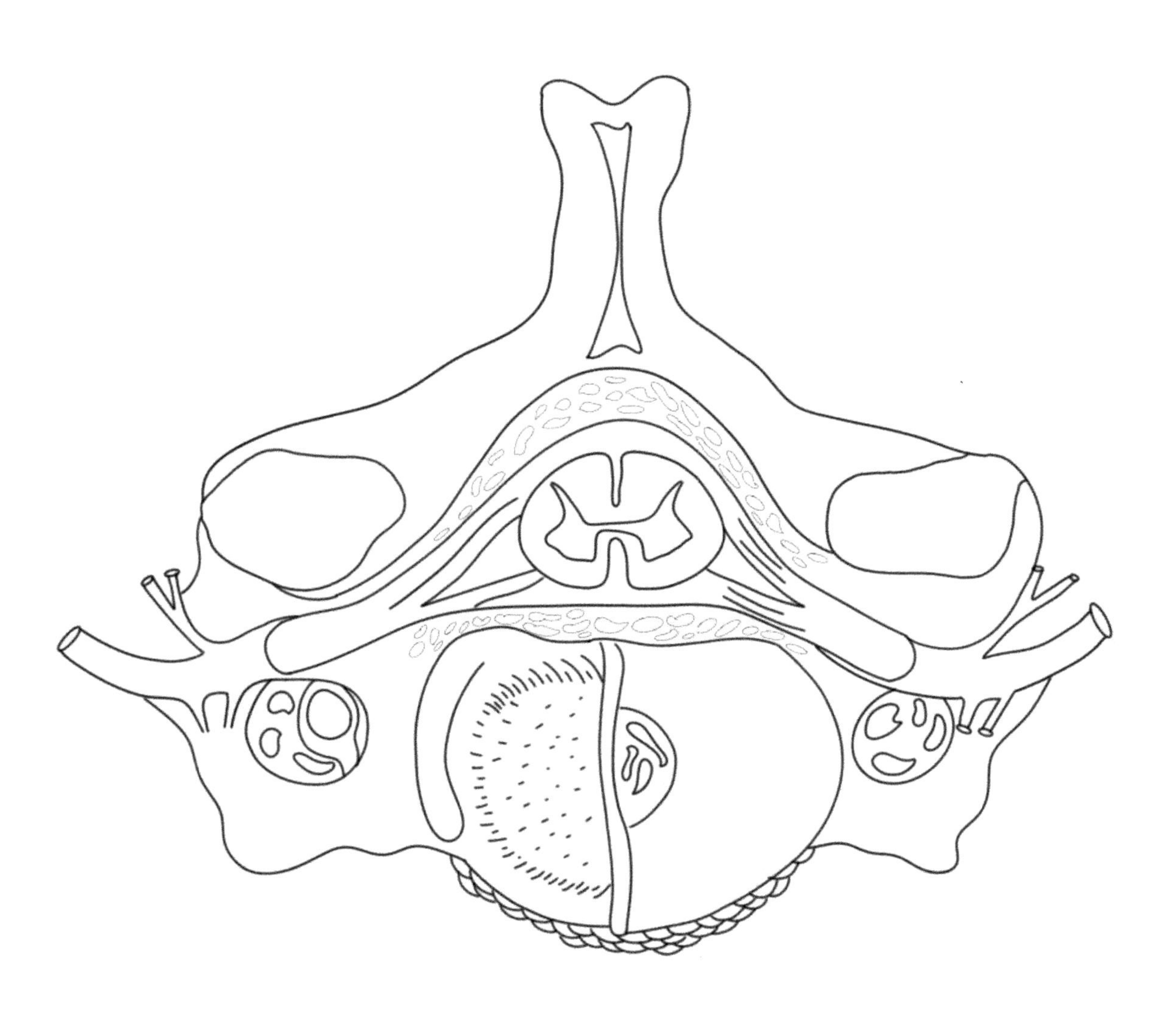

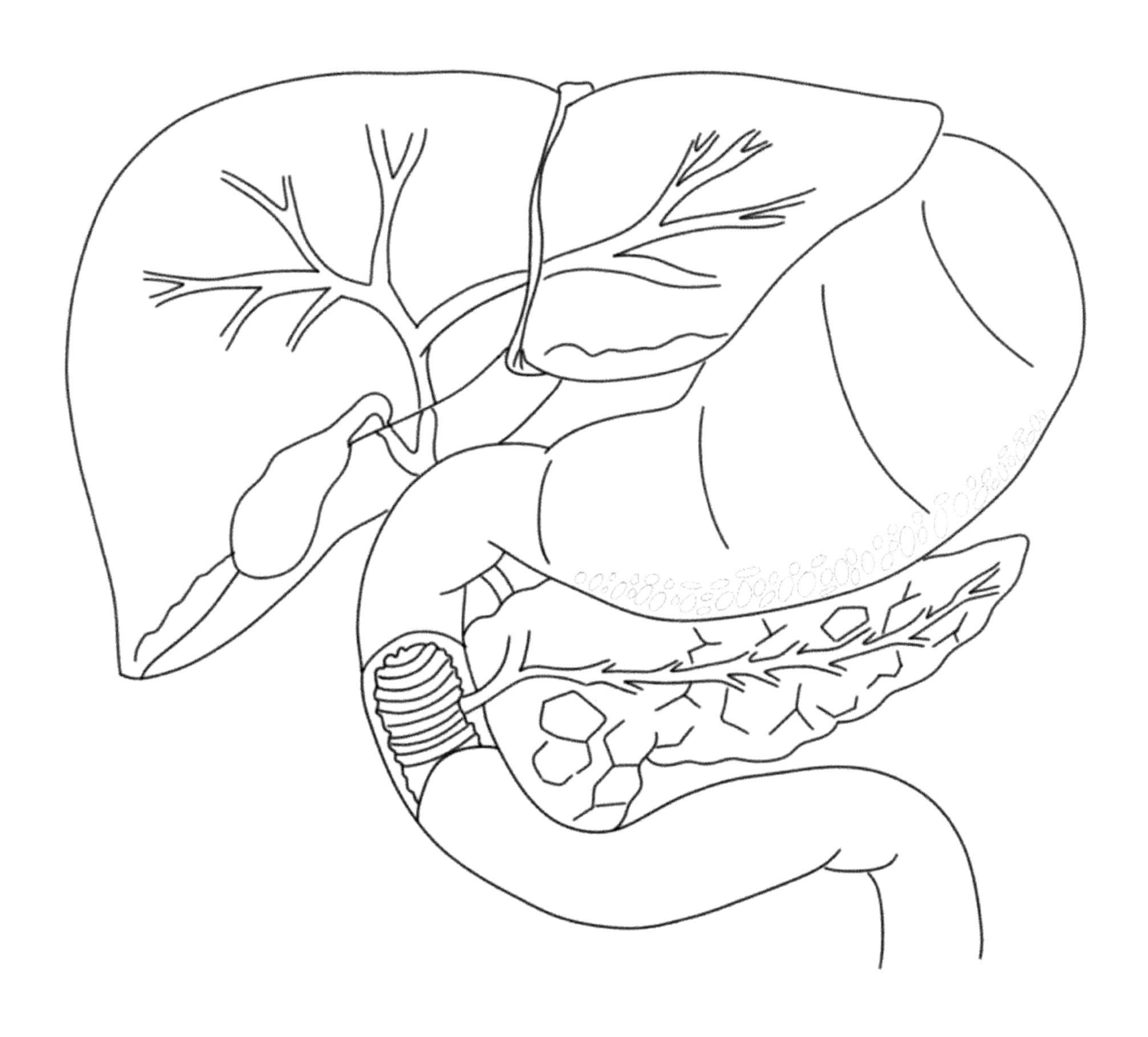

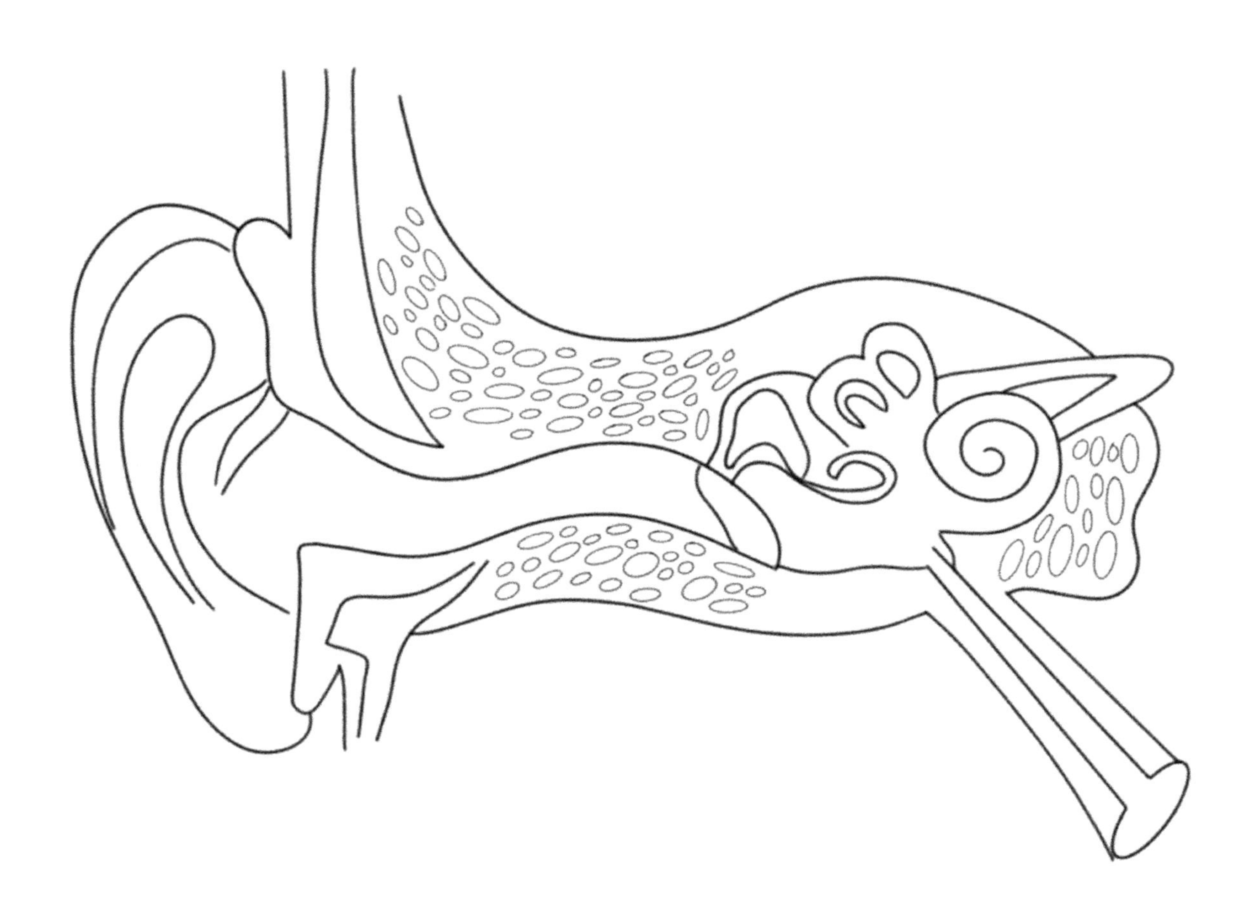

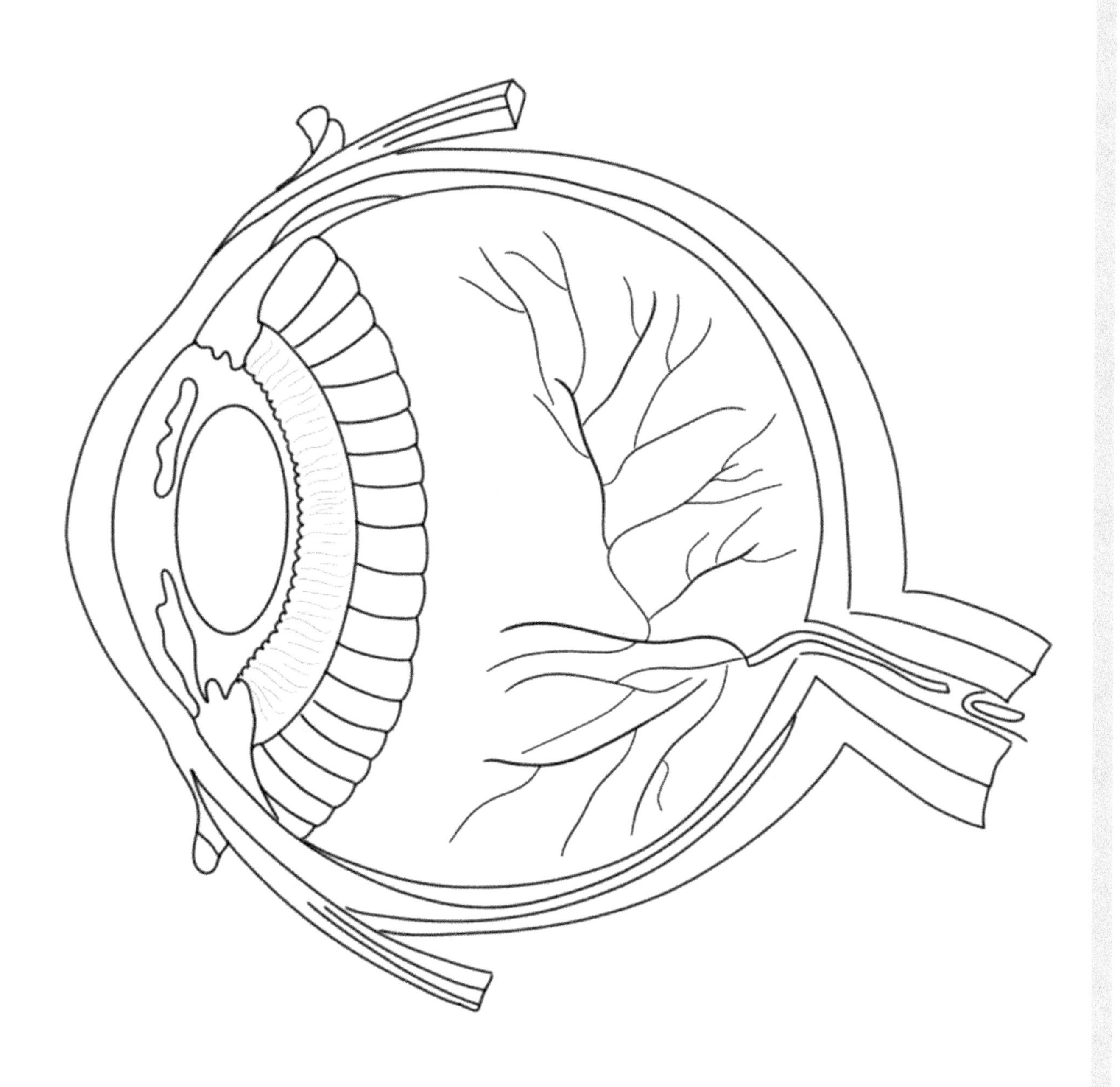

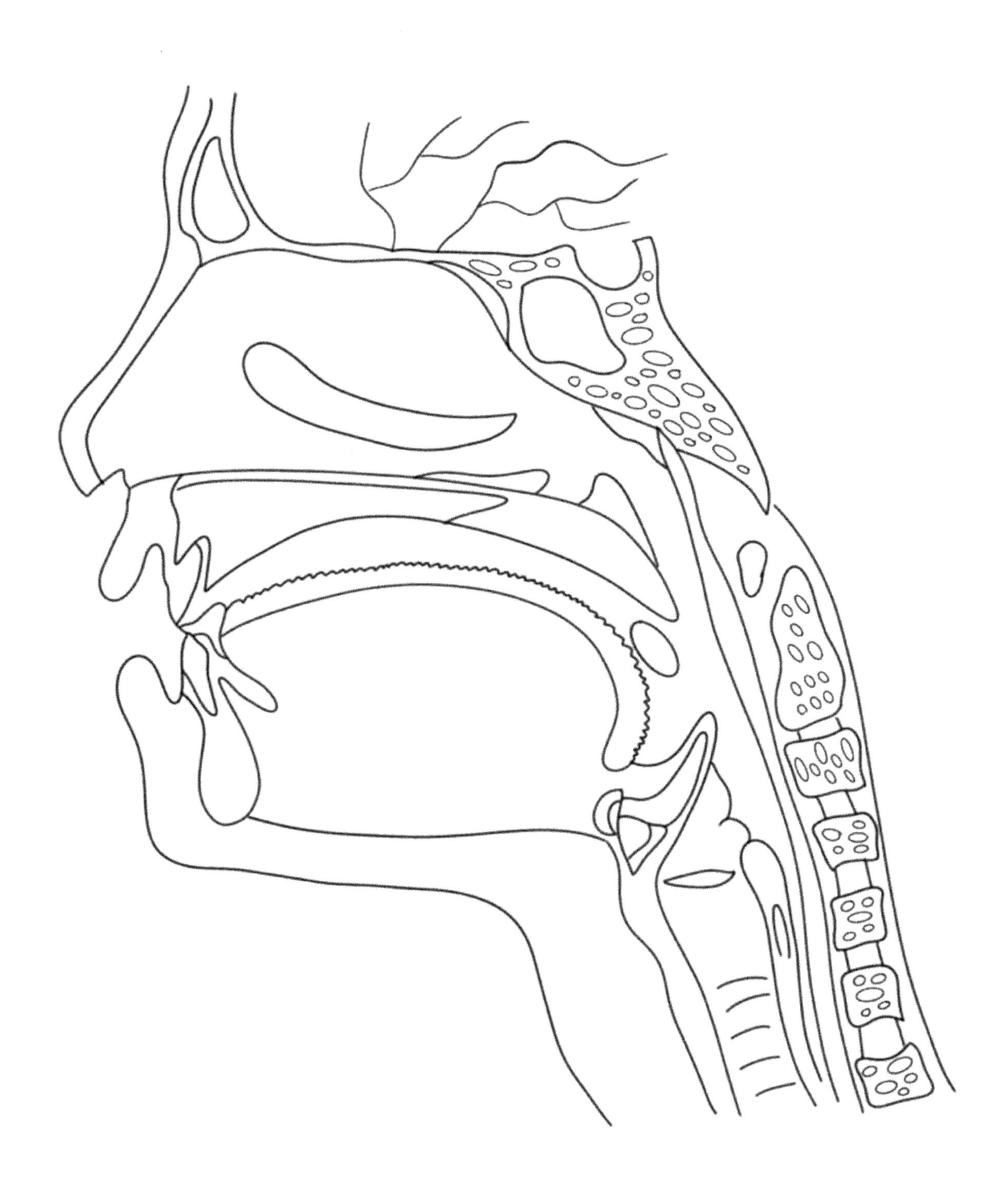

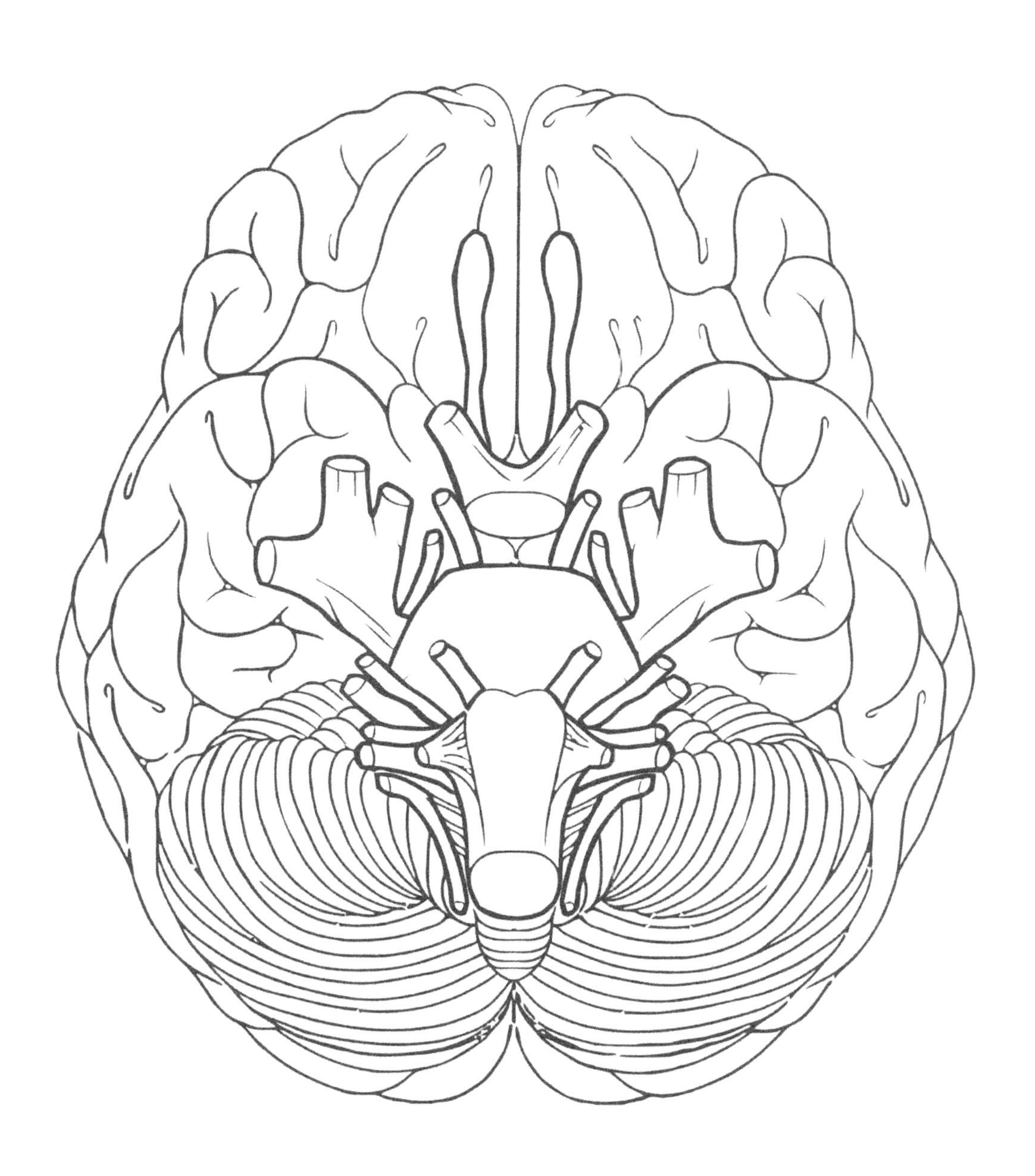

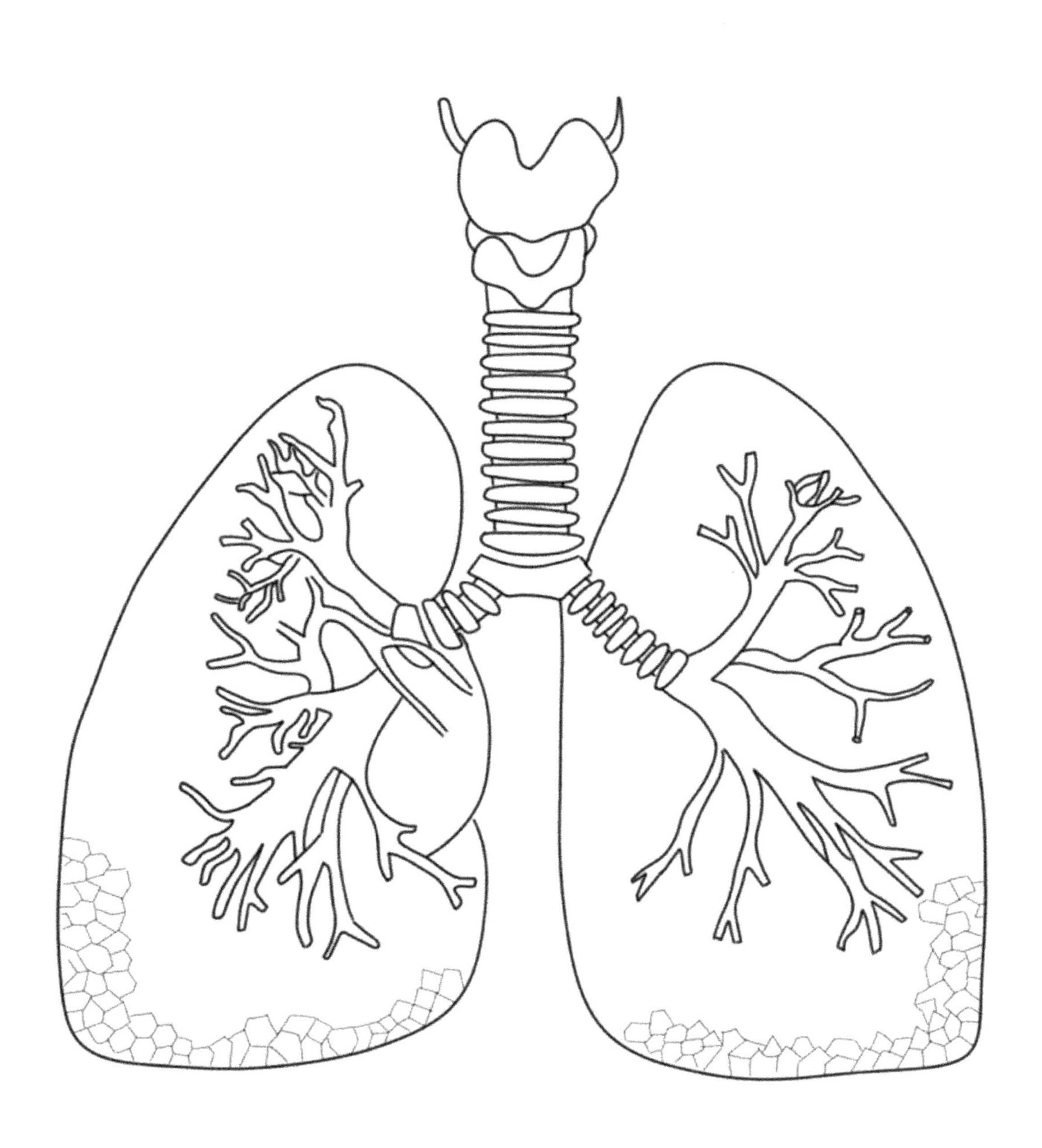

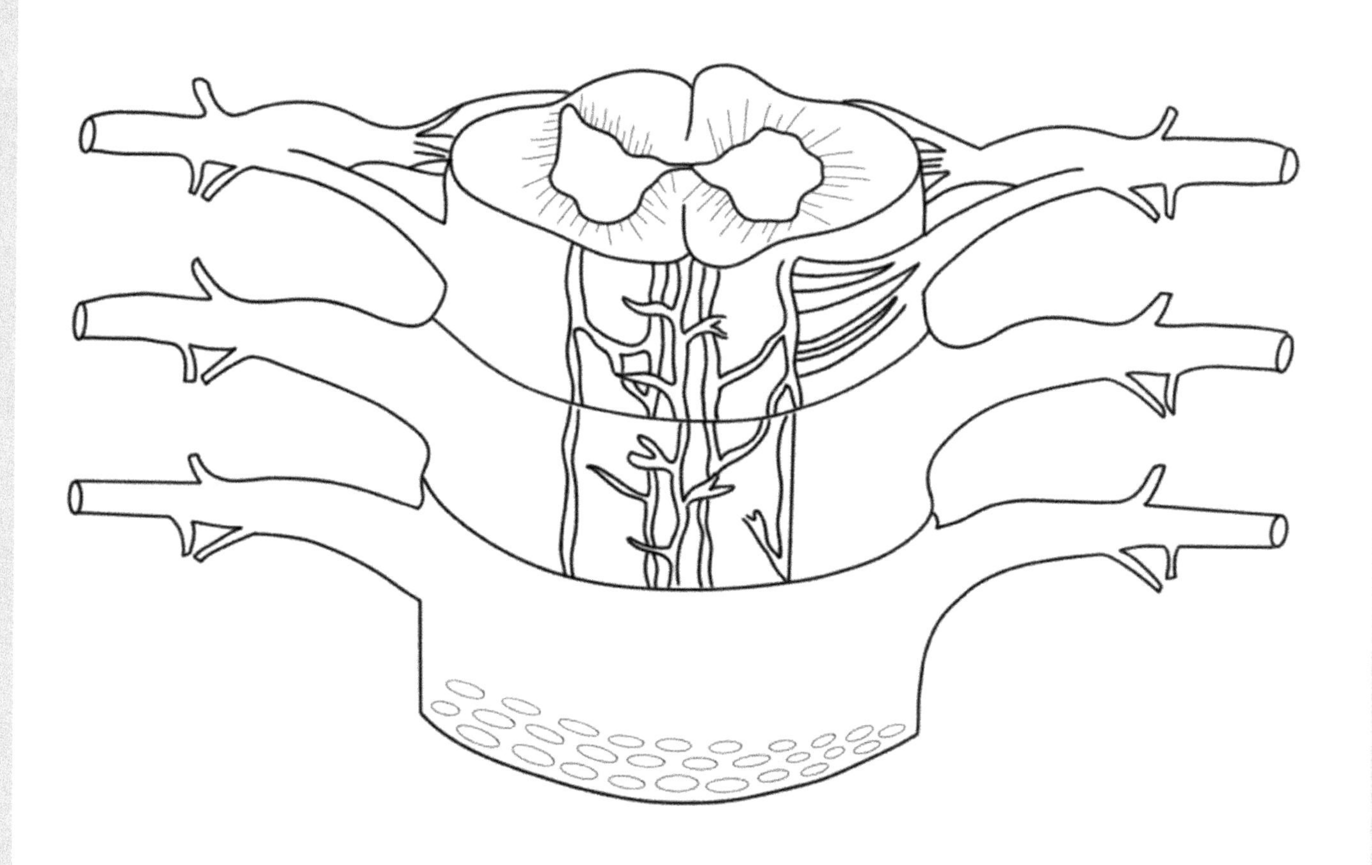

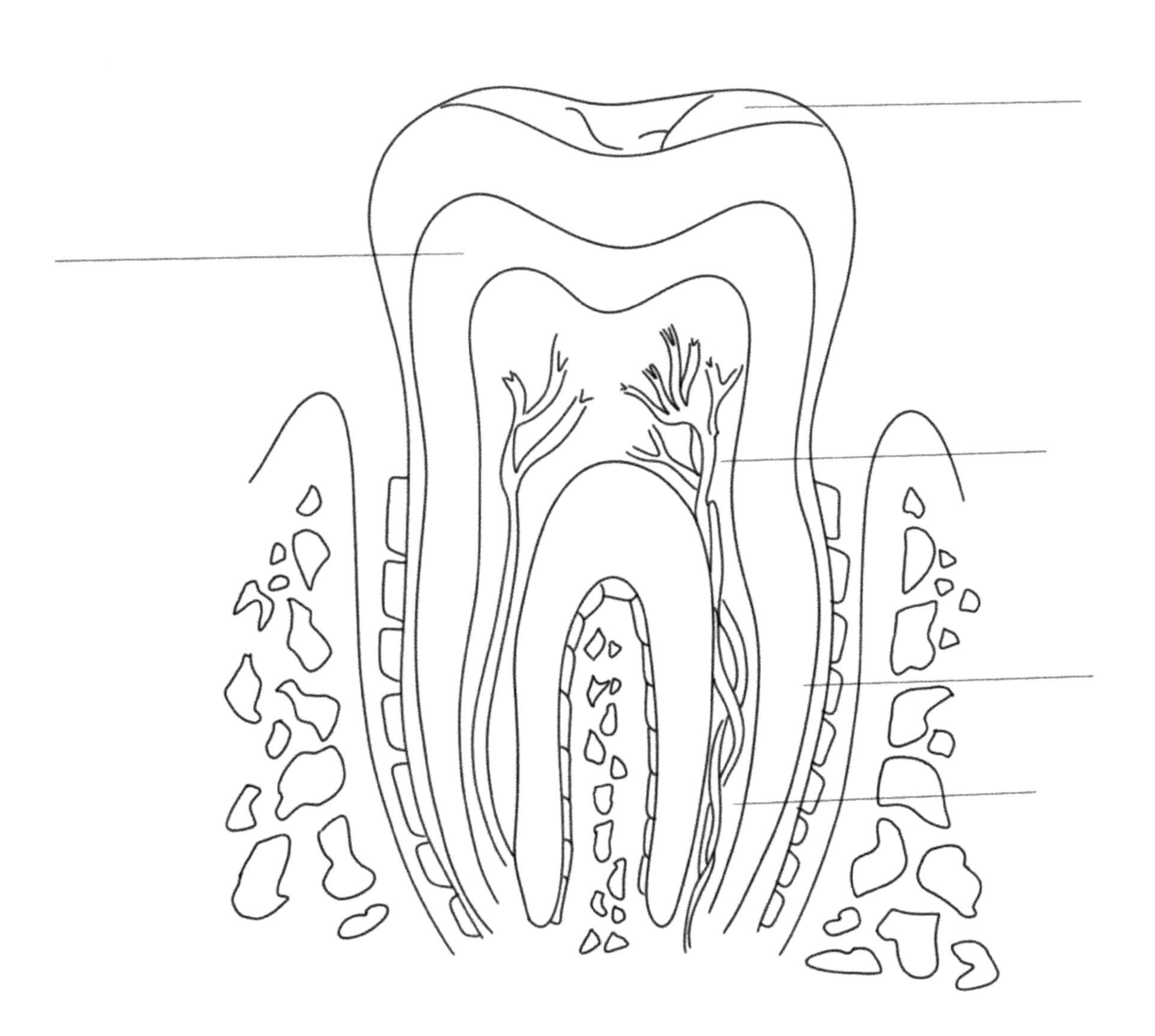

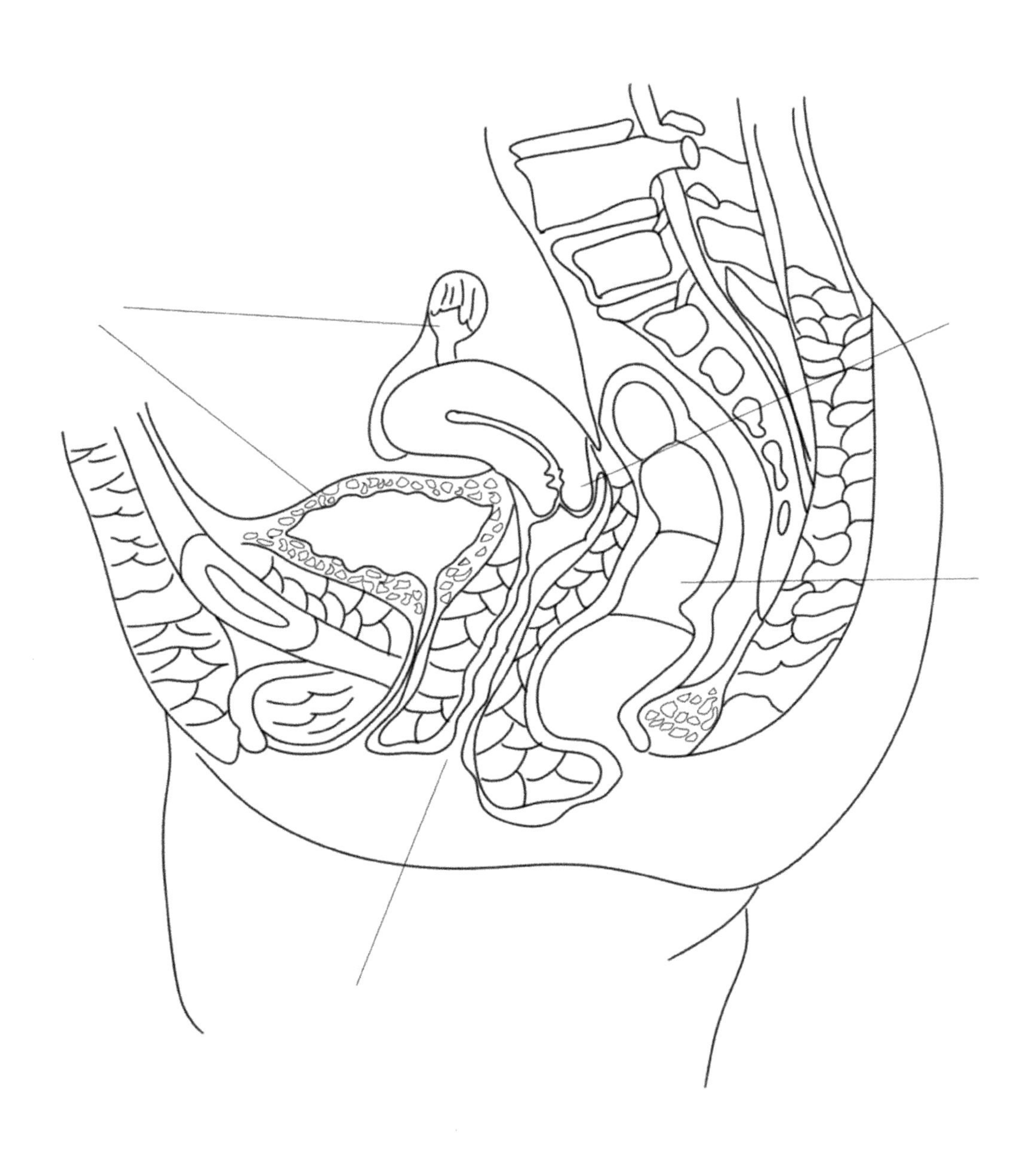

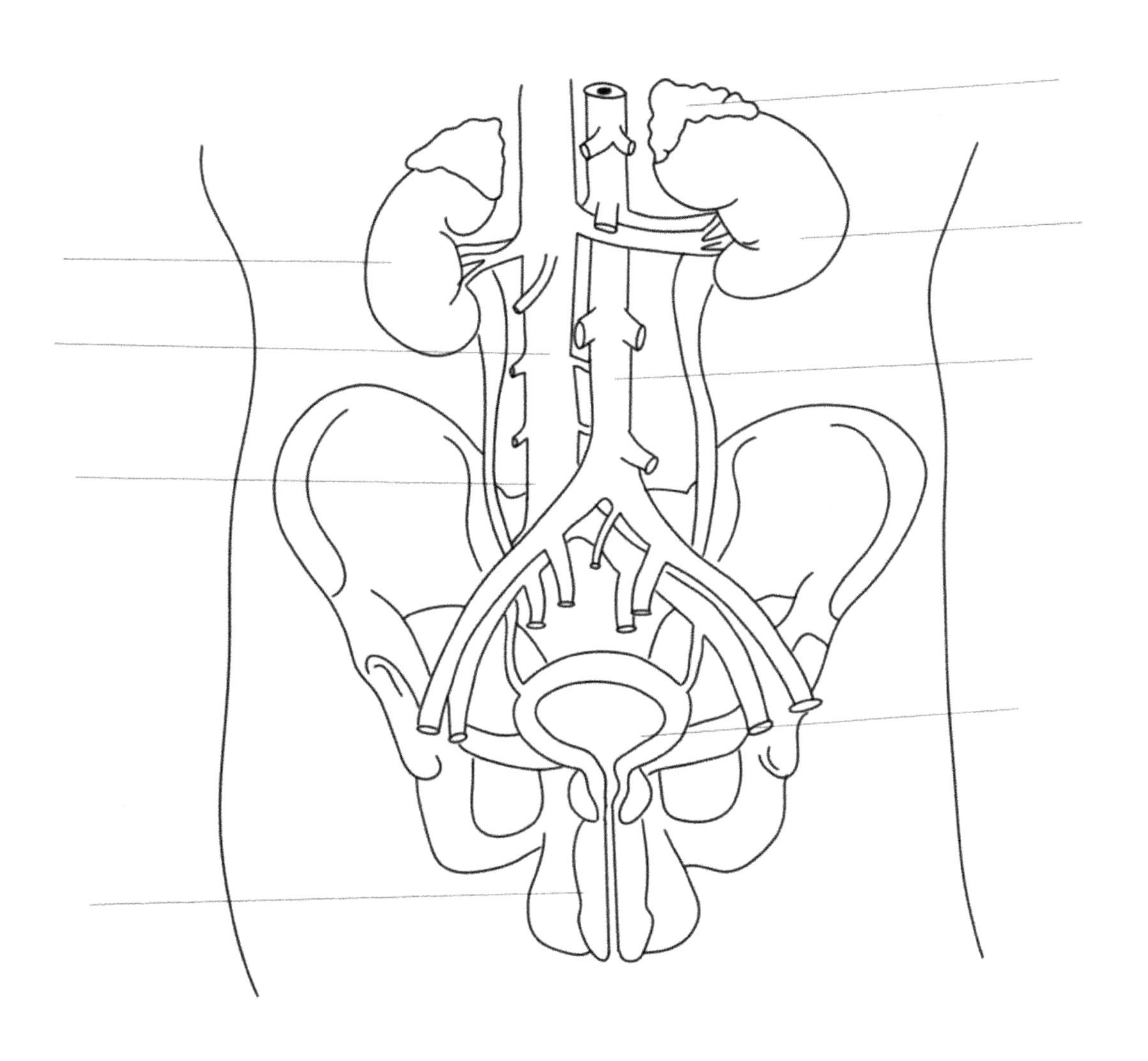

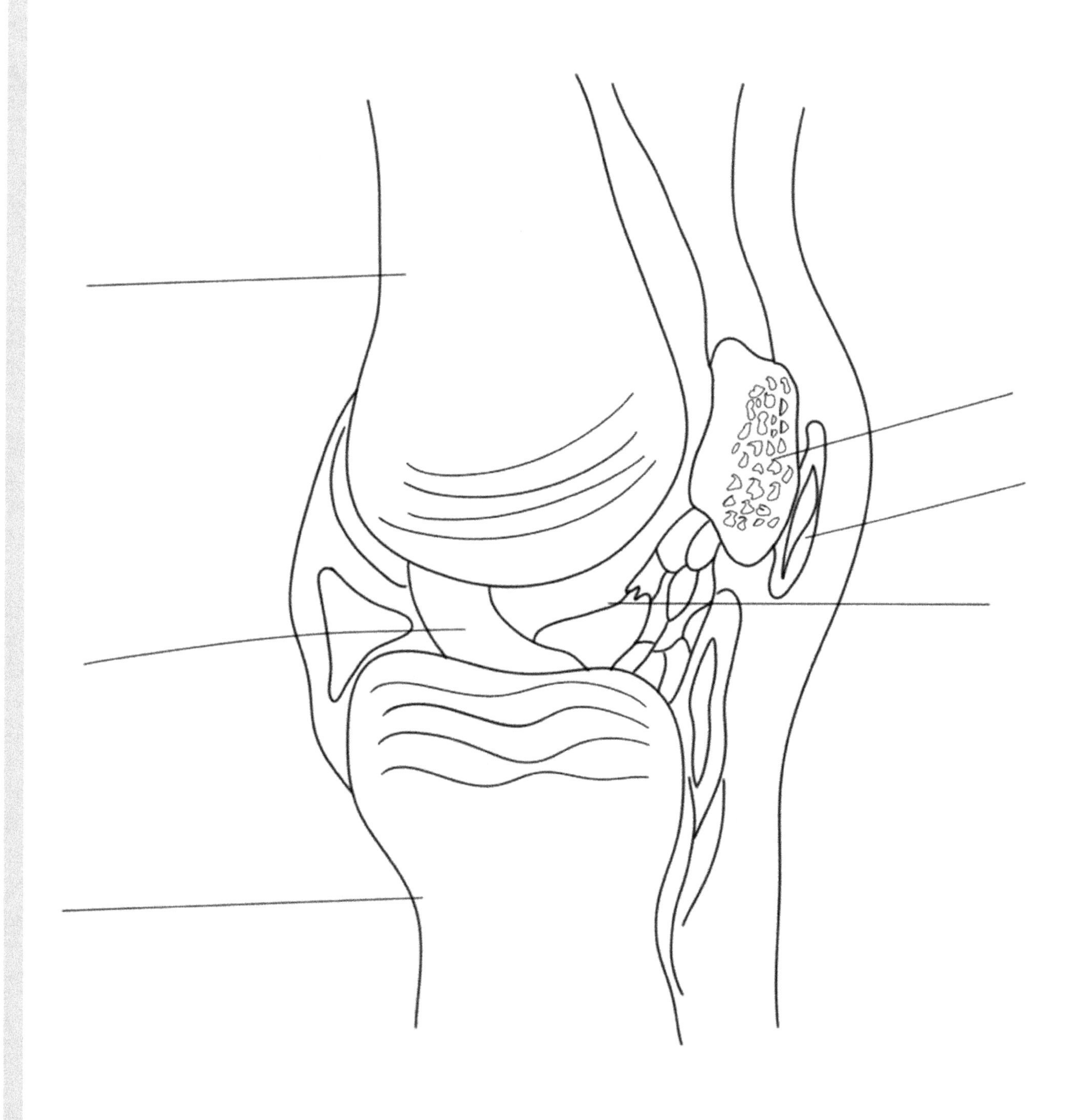

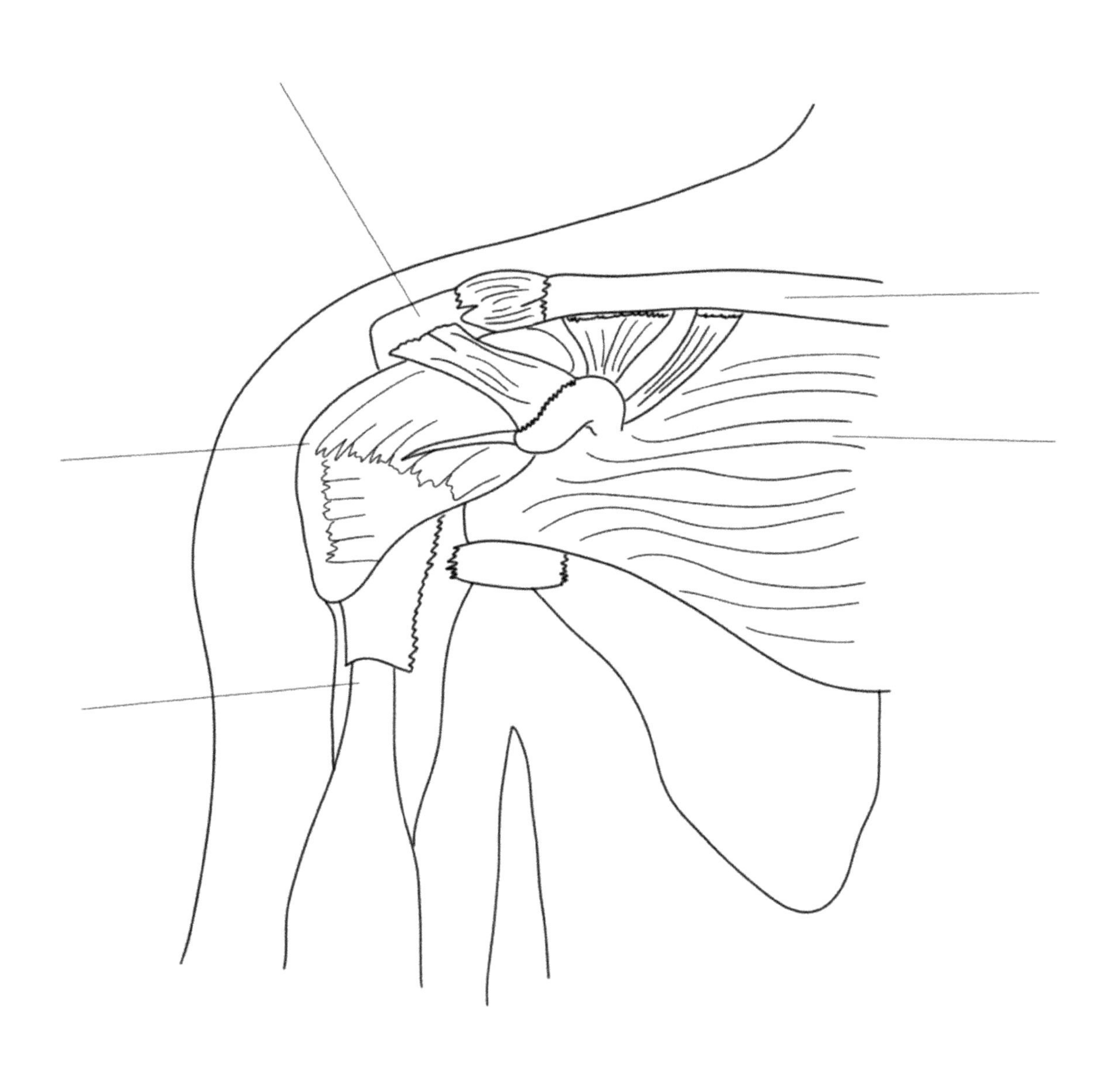

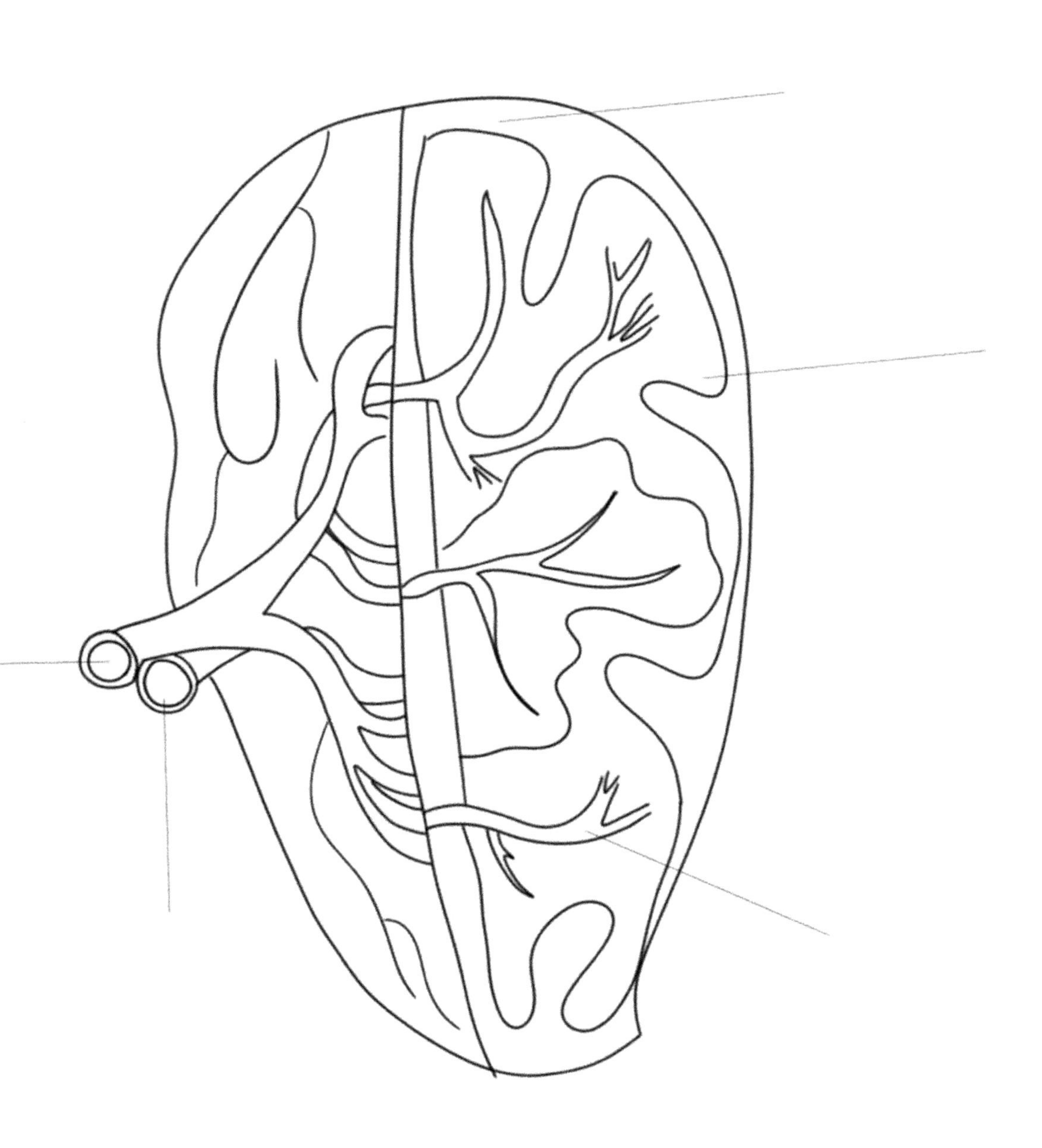

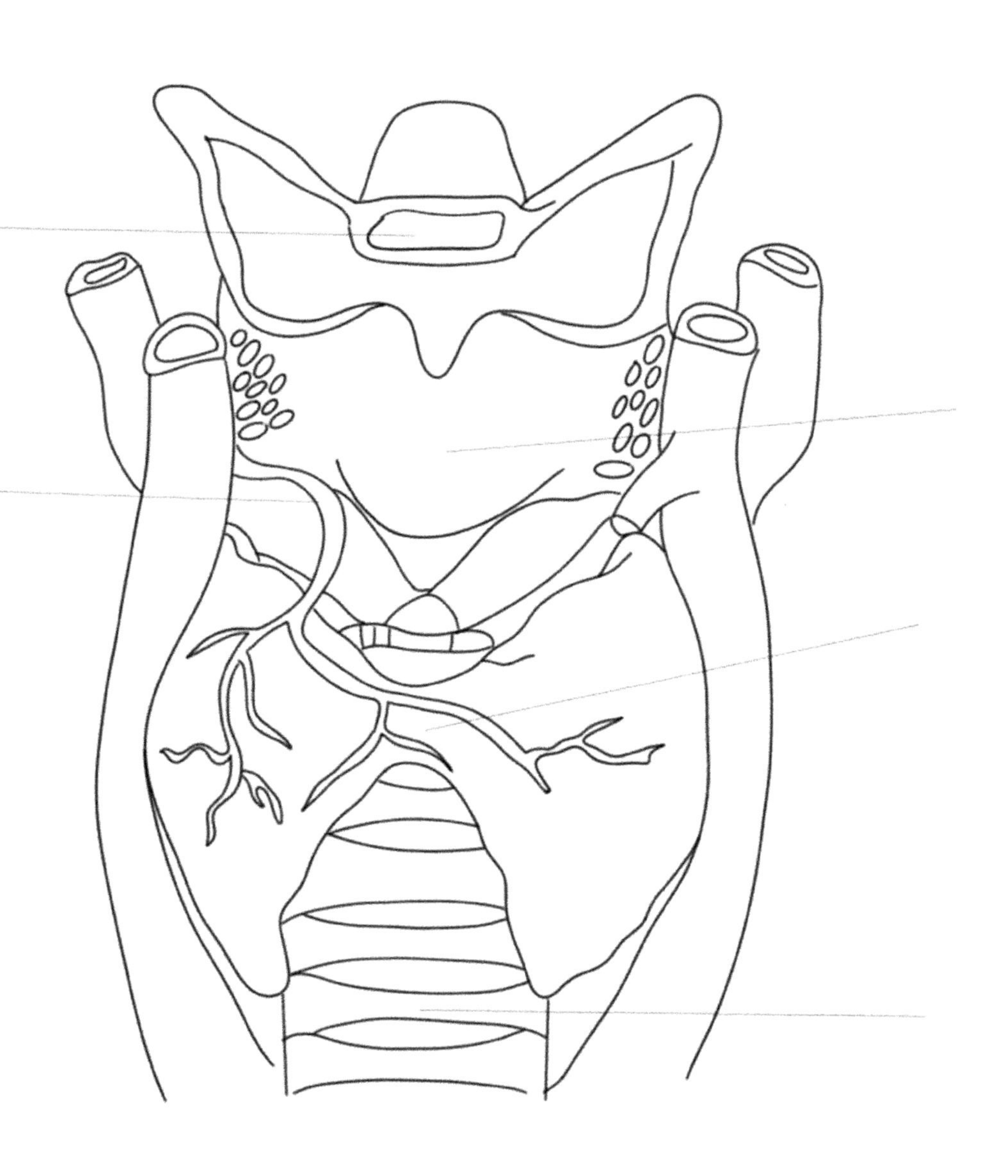

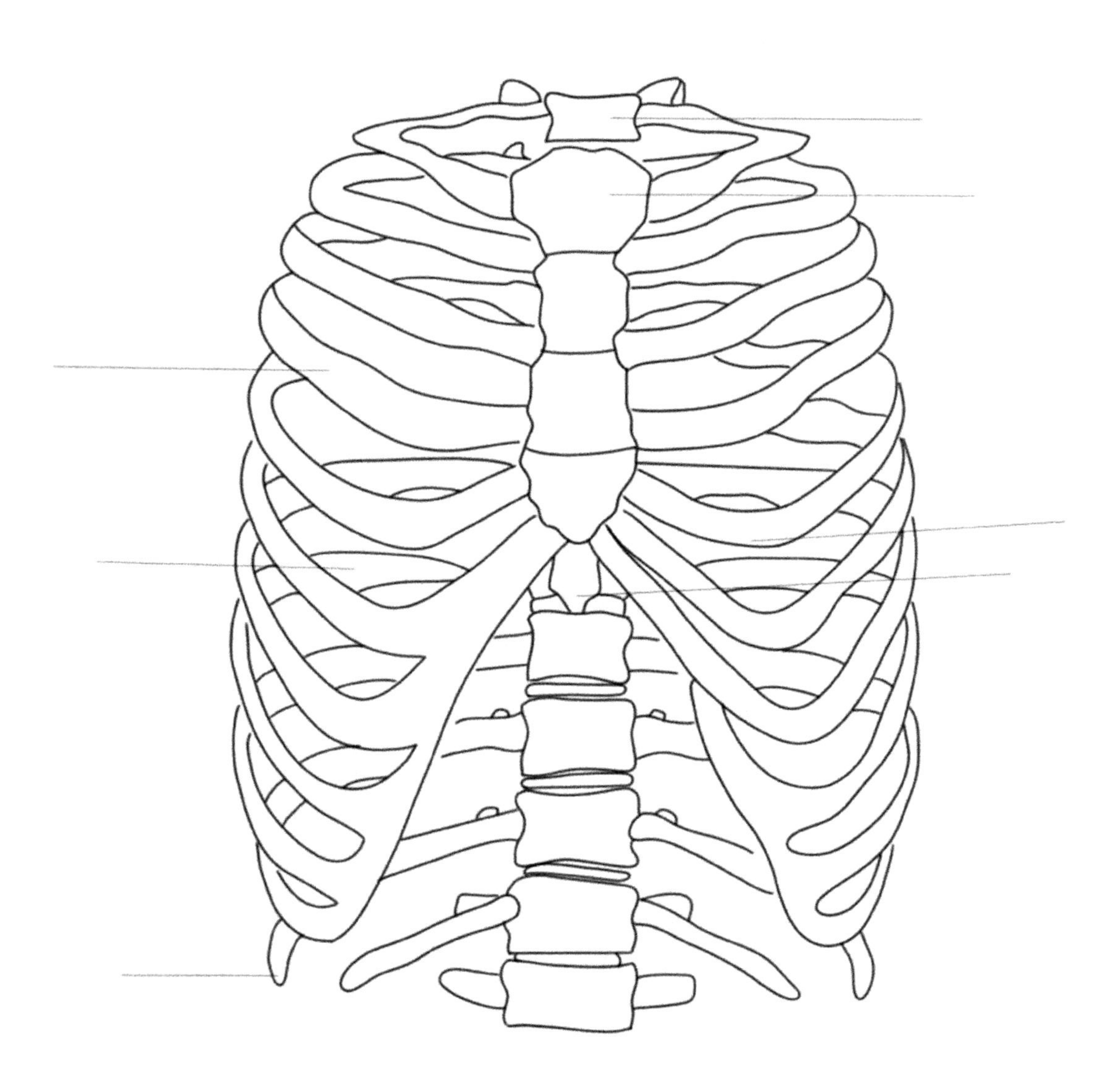

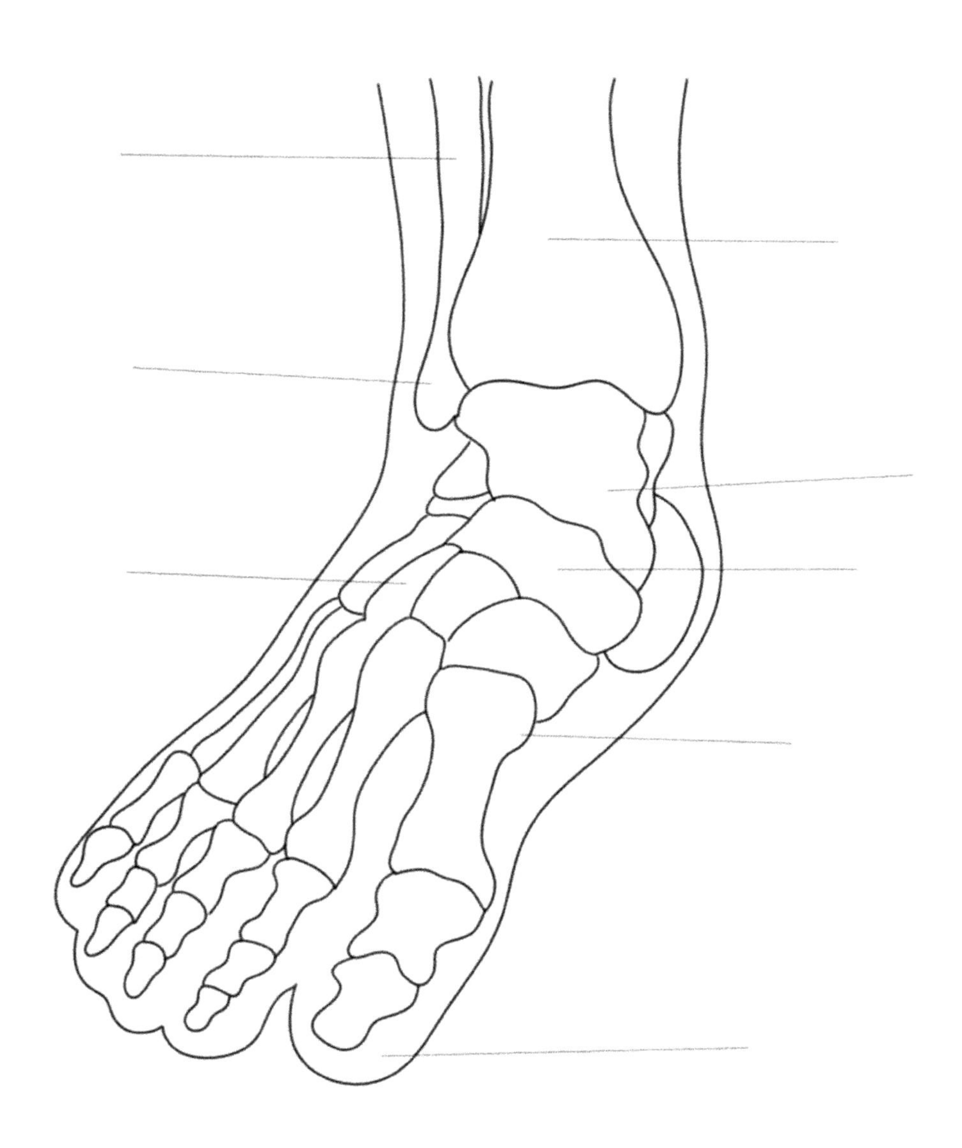

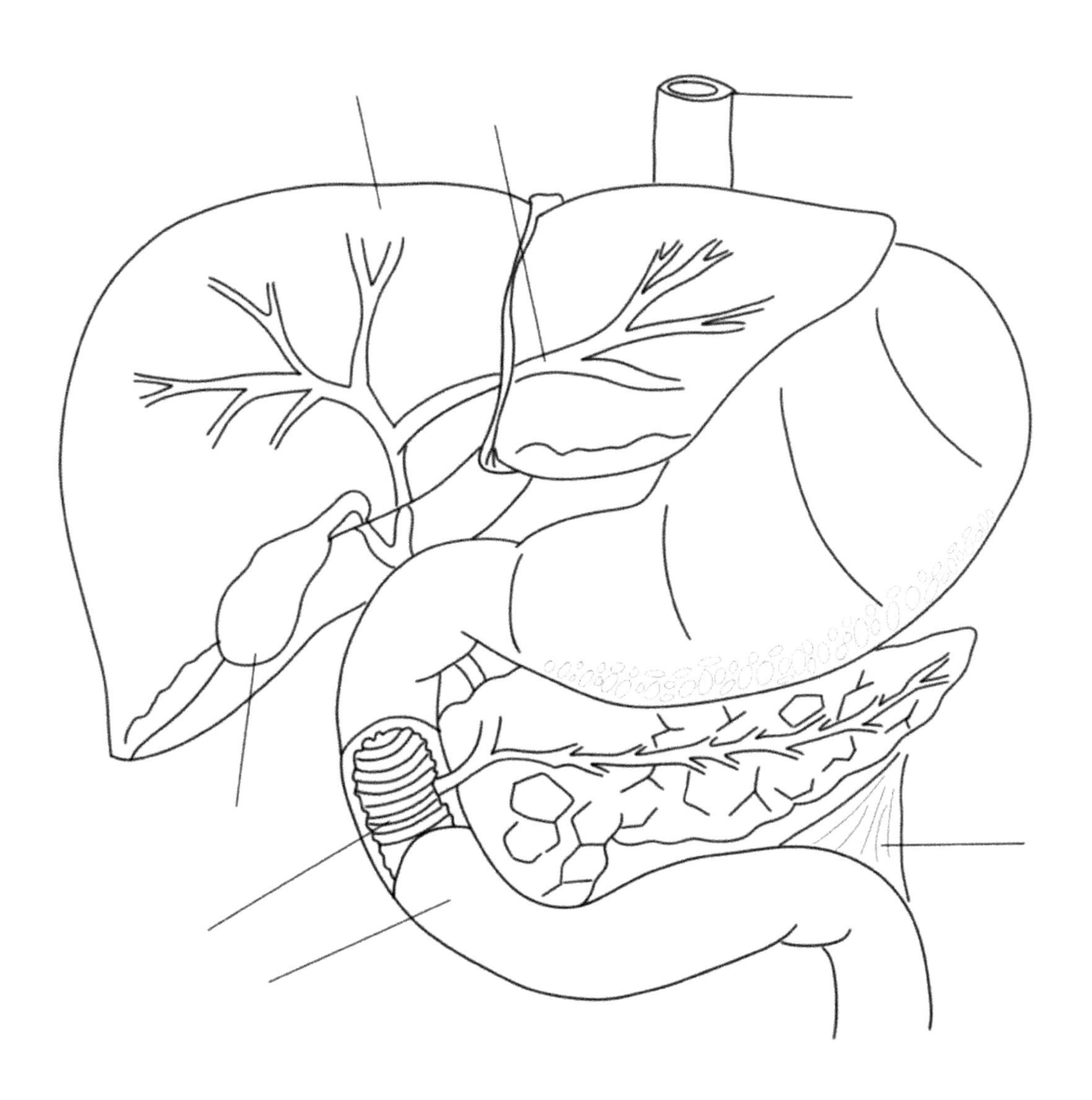

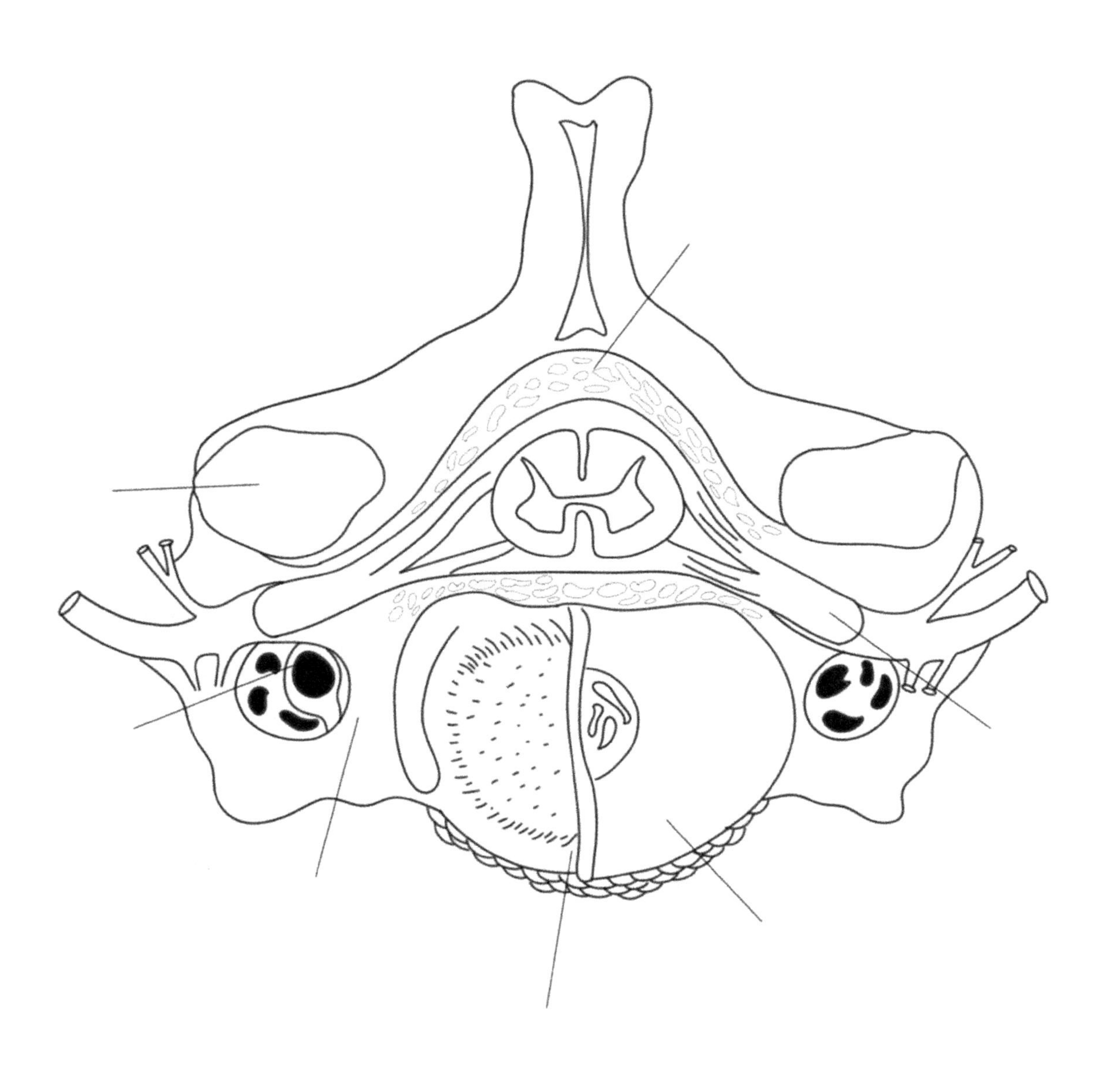

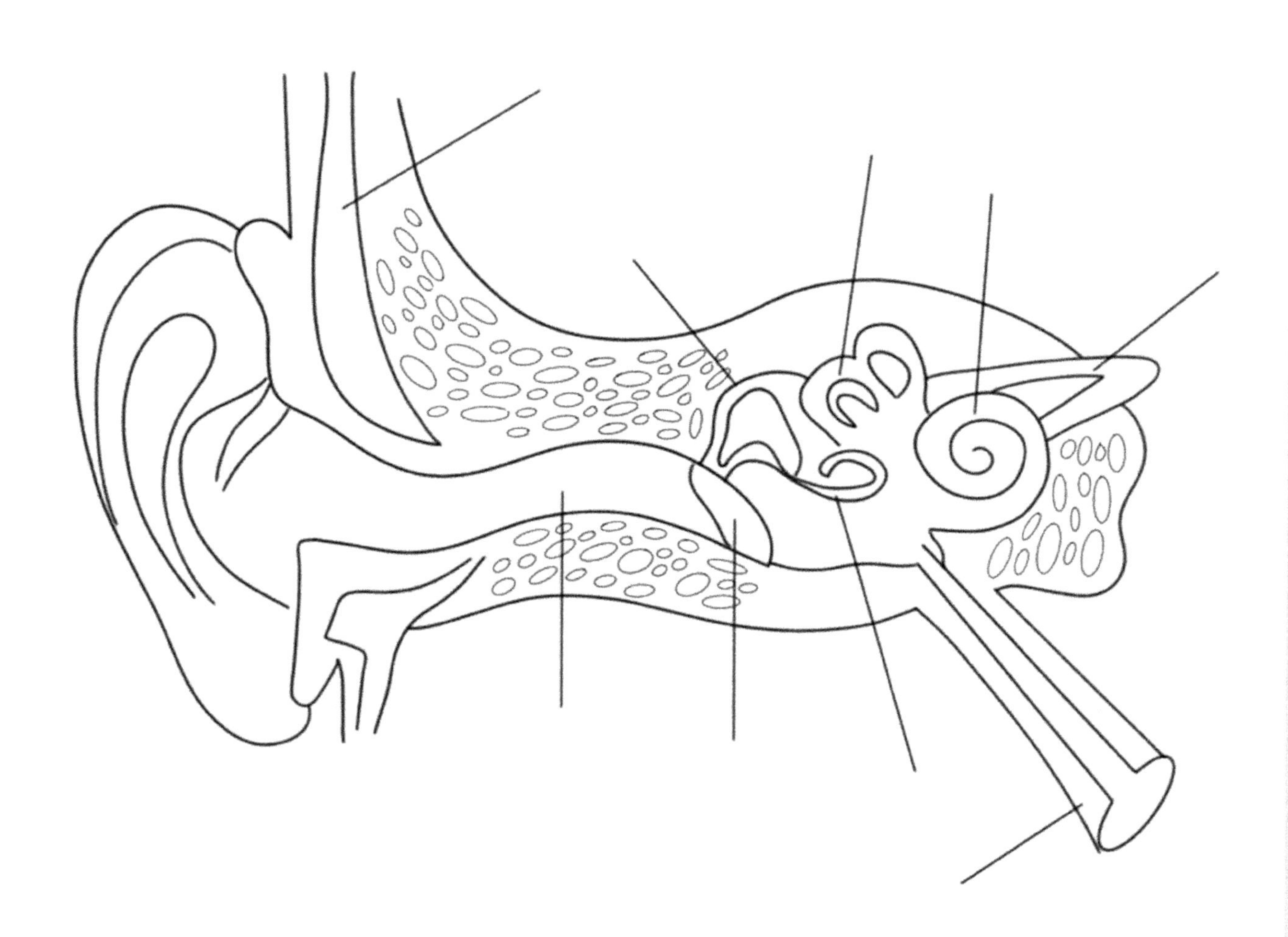

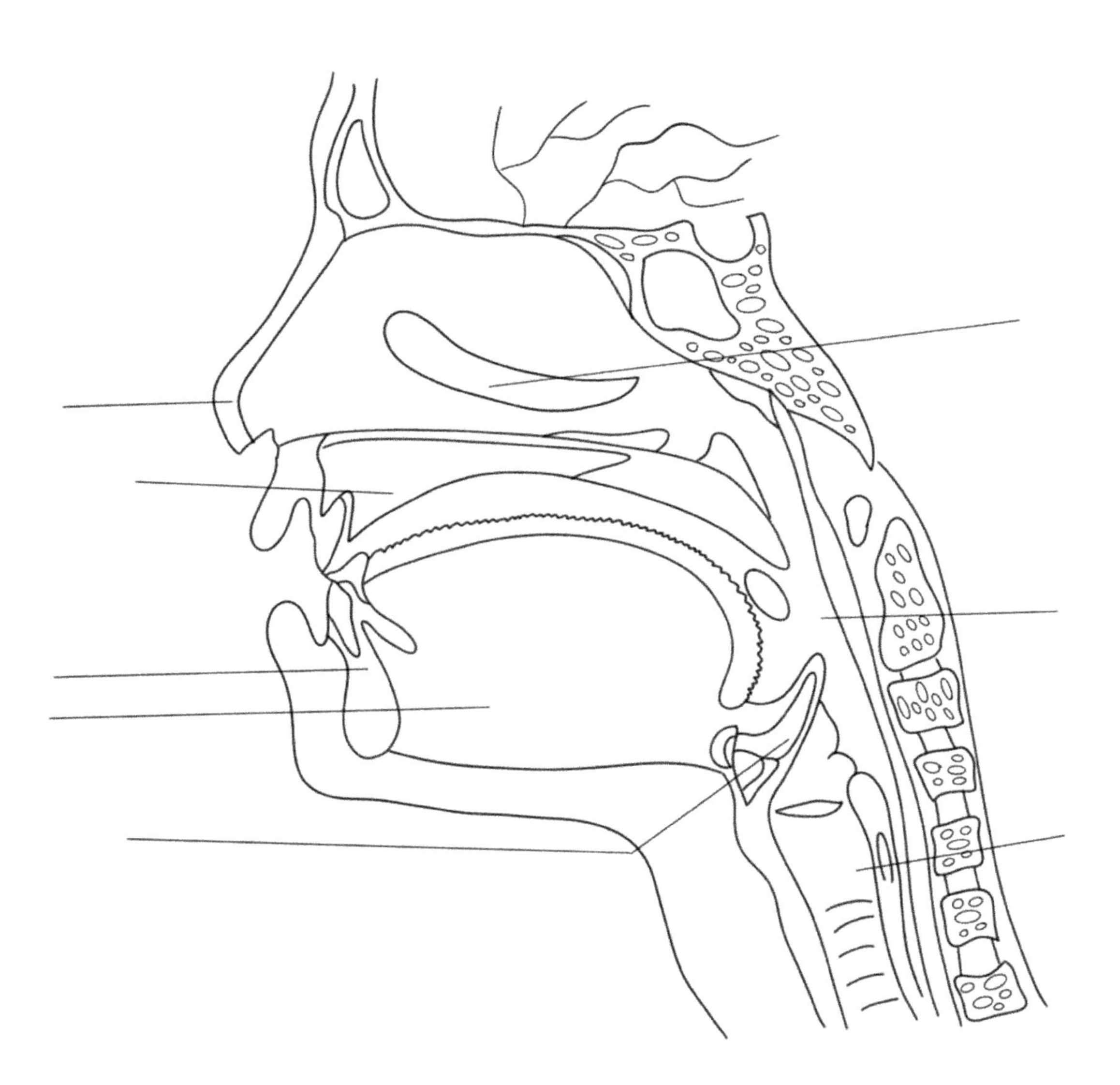

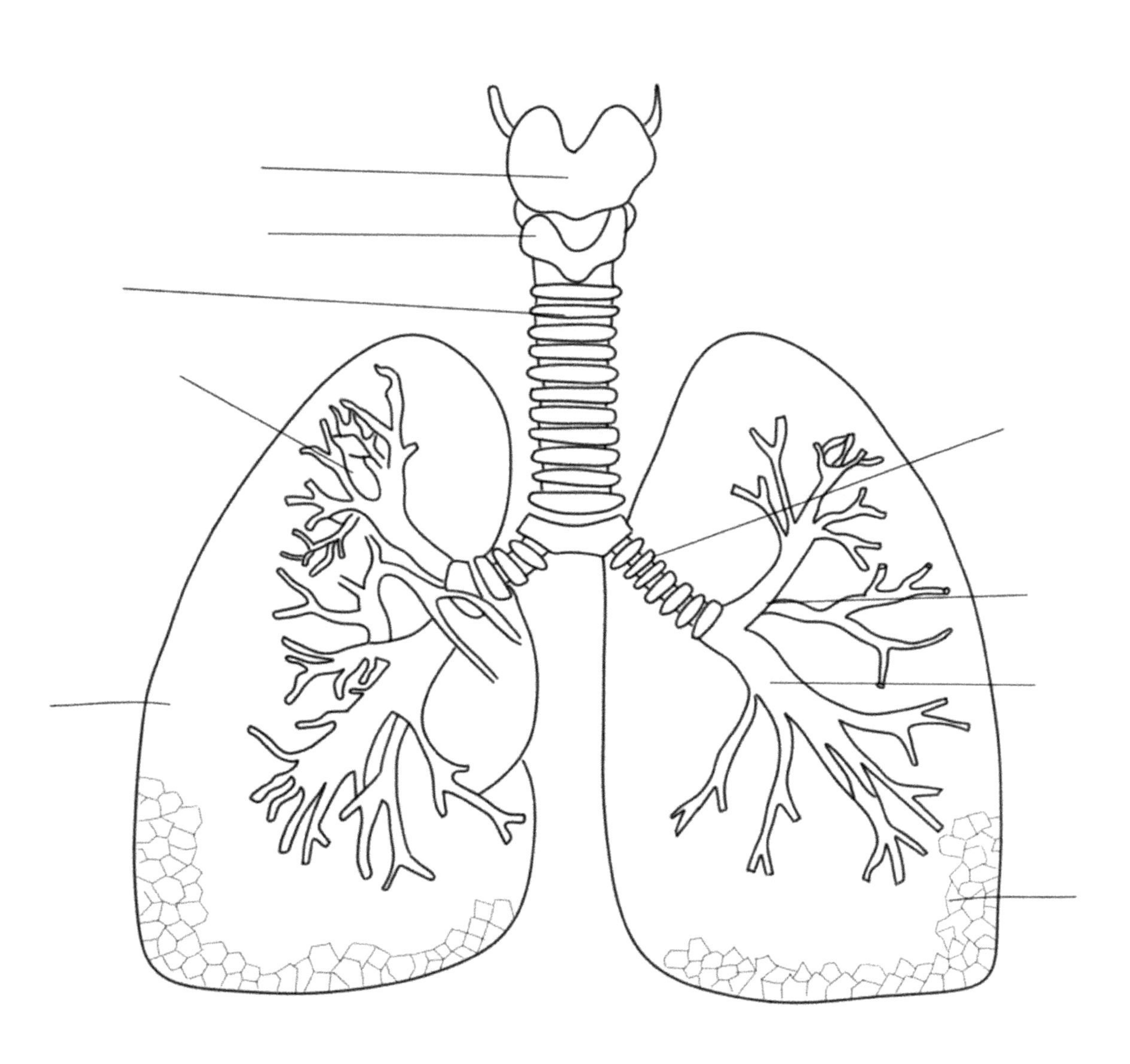

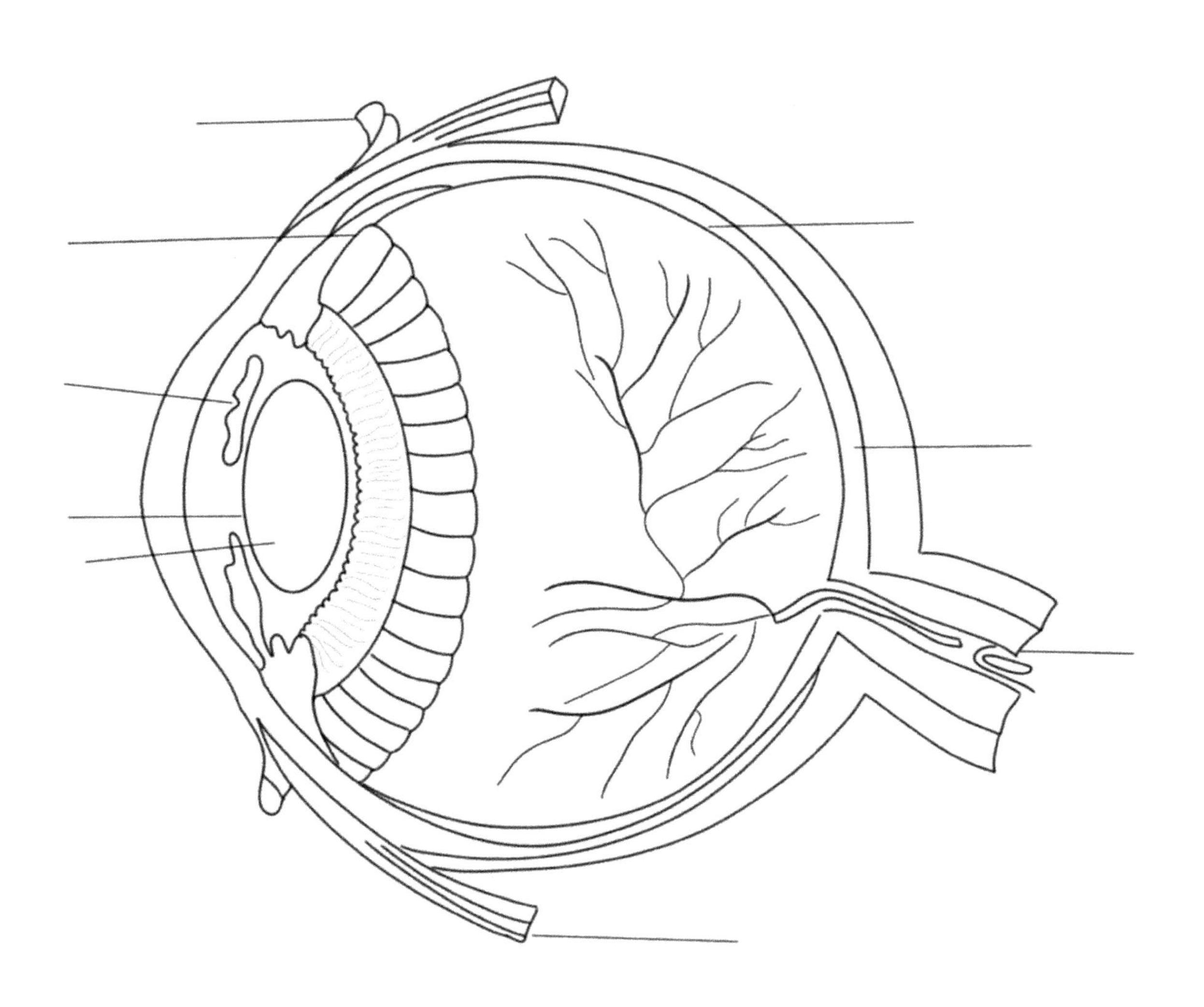

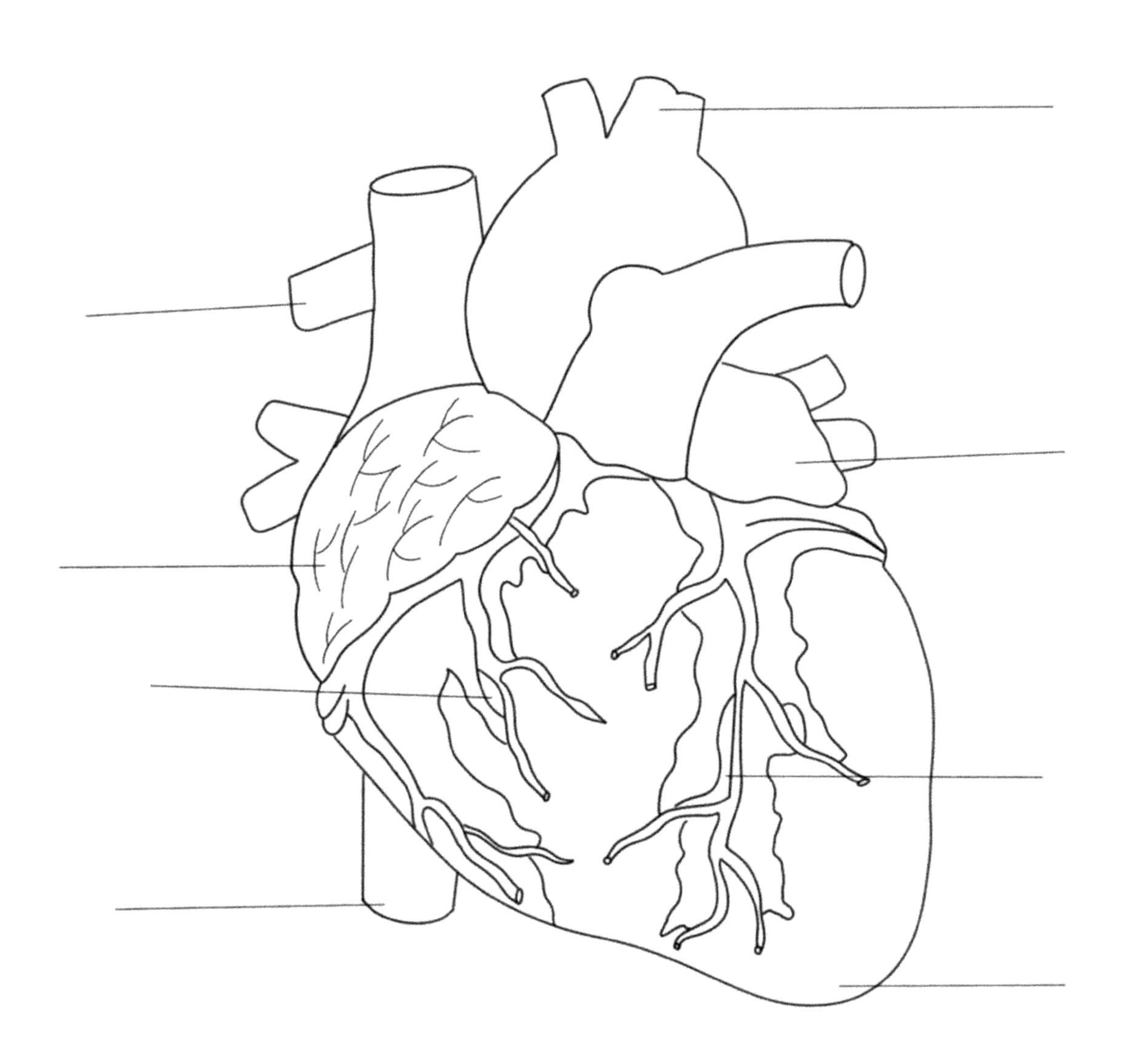

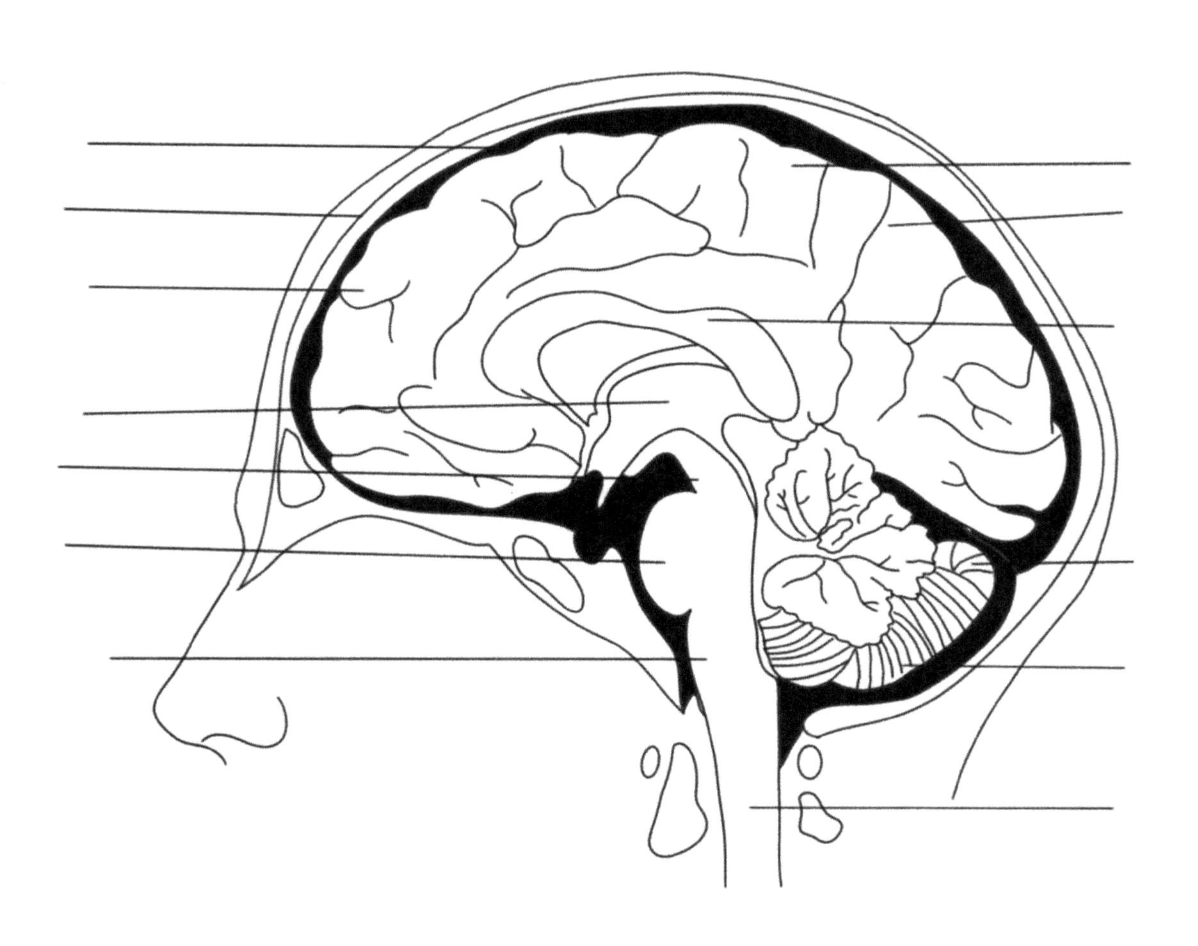

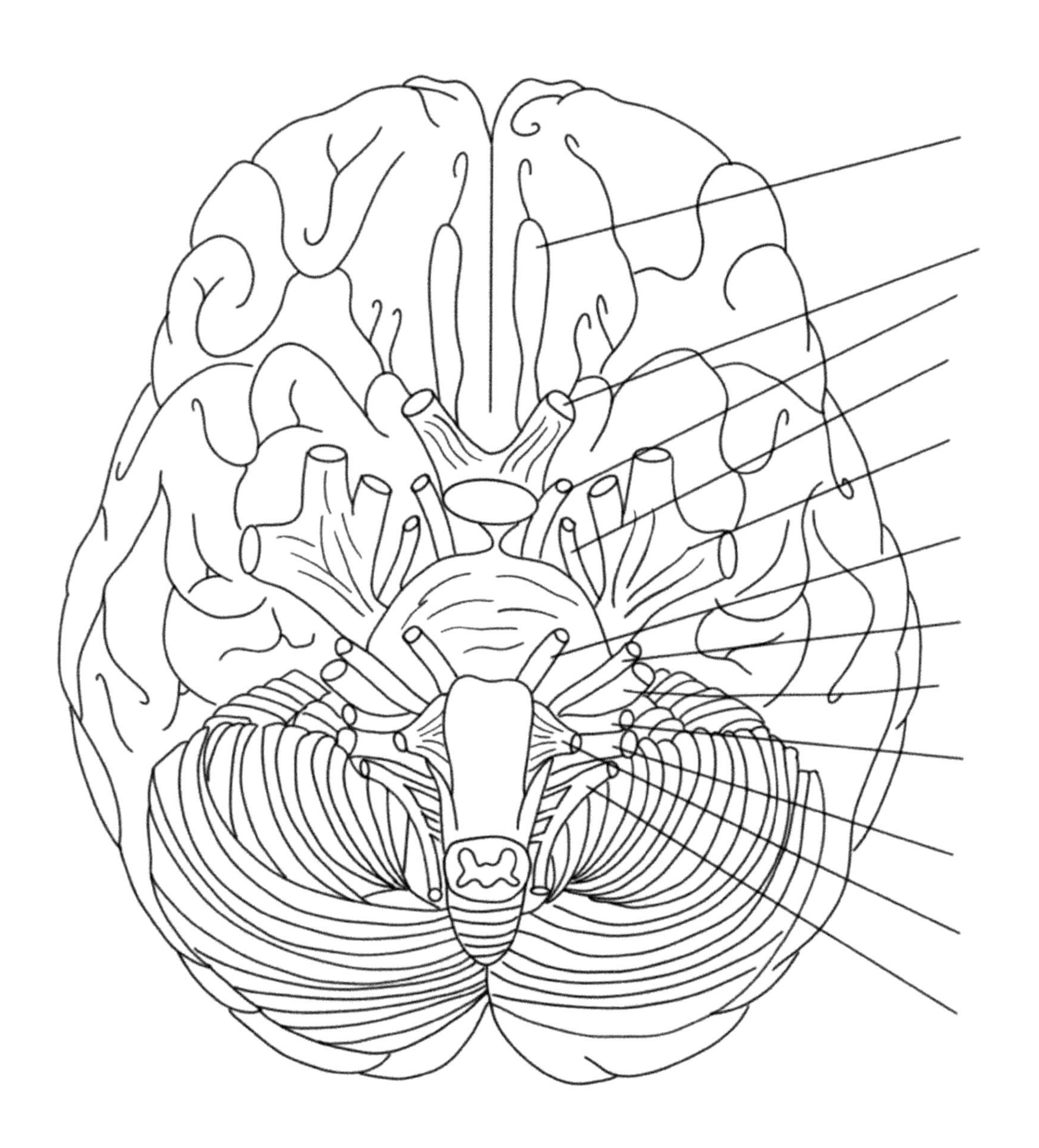

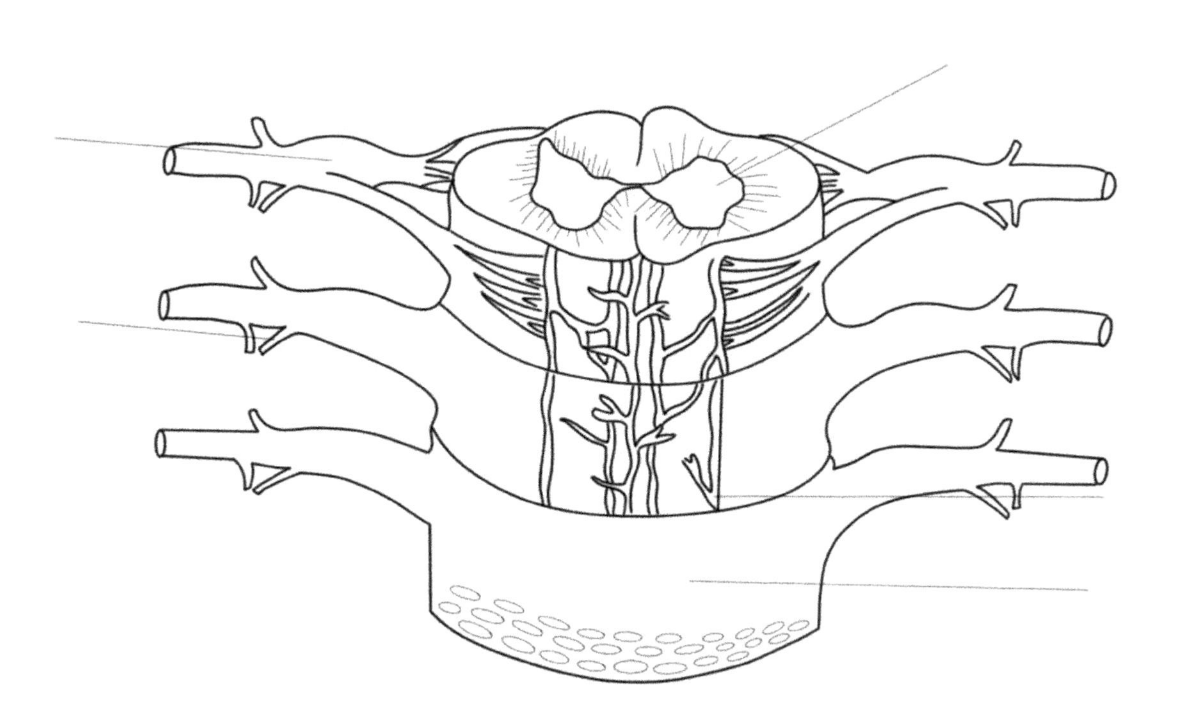

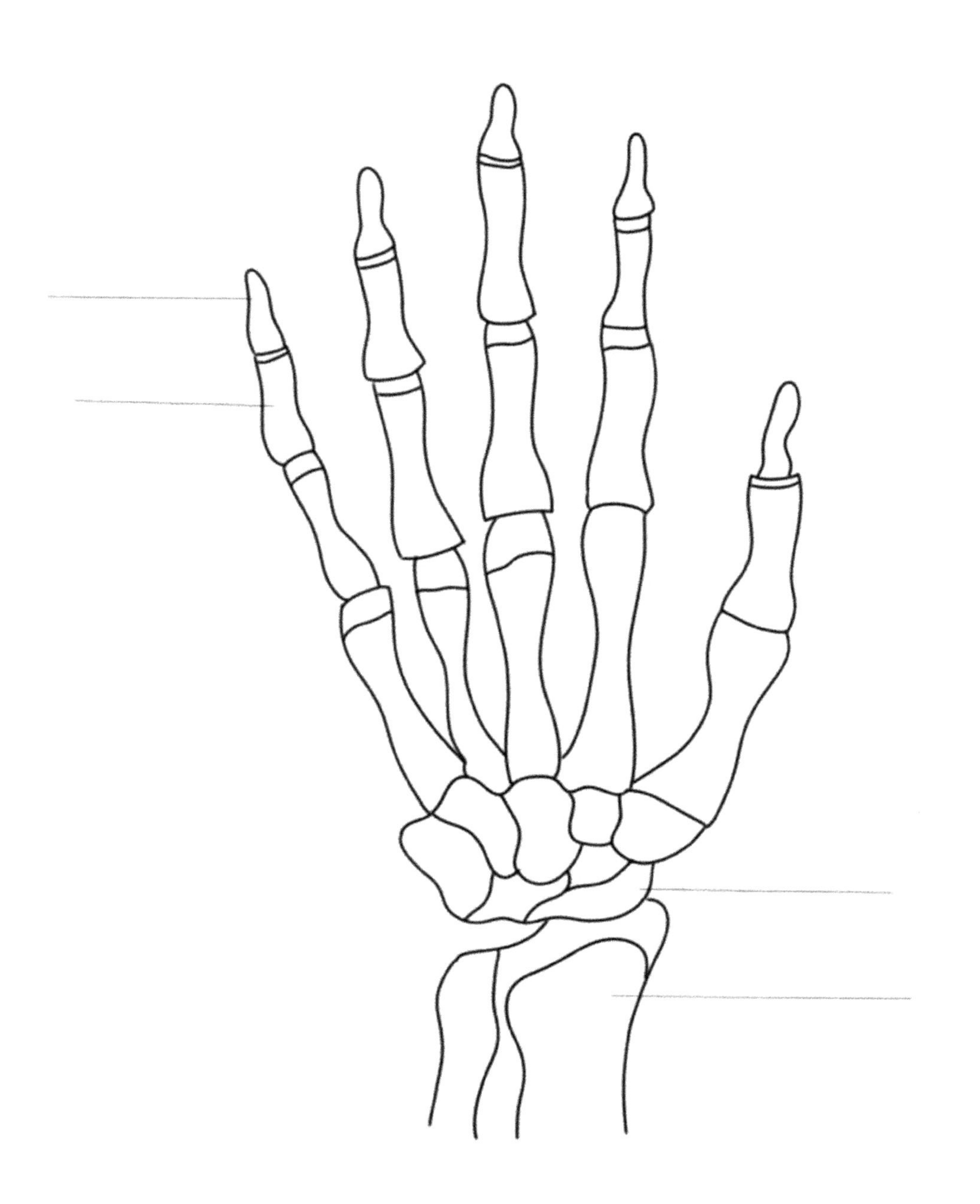